Synthetic Data for Deep Learning

Generate Synthetic Data for Decision Making and Applications with Python and R

Necmi Gürsakal
Sadullah Çelik
Esma Birişçi

Apress®

Synthetic Data for Deep Learning: Generate Synthetic Data for Decision Making and Applications with Python and R

Necmi Gürsakal
Bursa, Turkey

Sadullah Çelik
Aydın, Turkey

Esma Birişçi
Bursa, Turkey

ISBN-13 (pbk): 978-1-4842-8586-2
ISBN-13 (electronic): 978-1-4842-8587-9
https://doi.org/10.1007/978-1-4842-8587-9

Managing Director, Apress Media LLC: Welmoed Spahr
Acquisitions Editor: Celestin Suresh John
Development Editor: Laura Berendson
Coordinating Editor: Mark Powers

Cover designed by eStudioCalamar

Cover image by Simon Lee on Unsplash (www.unsplash.com)

Distributed to the book trade worldwide by Apress Media, LLC, 1 New York Plaza, New York, NY 10004, U.S.A. Phone 1-800-SPRINGER, fax (201) 348-4505, e-mail orders-ny@springer-sbm.com, or visit www.springeronline.com. Apress Media, LLC is a California LLC and the sole member (owner) is Springer Science + Business Media Finance Inc (SSBM Finance Inc). SSBM Finance Inc is a **Delaware** corporation.

For information on translations, please e-mail booktranslations@springernature.com; for reprint, paperback, or audio rights, please e-mail bookpermissions@springernature.com.

Apress titles may be purchased in bulk for academic, corporate, or promotional use. eBook versions and licenses are also available for most titles. For more information, reference our Print and eBook Bulk Sales web page at http://www.apress.com/bulk-sales.

Any source code or other supplementary material referenced by the author in this book is available to readers on GitHub (https://github.com/Apress). For more detailed information, please visit http://www.apress.com/source-code.

Printed on acid-free paper

This book is dedicated to our mothers.

Table of Contents

About the Authors ix

About the Technical Reviewer xi

Preface xiii

Introduction xv

Chapter 1: An Introduction to Synthetic Data 1

What Synthetic Data is? 1

Why is Synthetic Data Important? 2

Synthetic Data for Data Science and Artificial Intelligence 3

Accuracy Problems 4

The Lifecycle of Data 5

Data Collection versus Privacy 7

Data Privacy and Synthetic Data 8

Synthetic Data and Data Quality 10

Aplications of Synthetic Data 10

Financial Services 11

Manufacturing 12

Healthcare 14

Automotive 16

Robotics 17

Security 18

Social Media 19

Marketing 20

Natural Language Processing 21

Computer Vision ... 22
Summary ... 27
References ... 27
Chapter 2: Foundations of Synthetic data ... 31
How to Generated Fair Synthetic Data? ... 31
Generating Synthetic Data in A Simple Way ... 32
Using Video Games to Create Synthetic Data ... 37
The Synthetic-to-Real Domain Gap ... 42
Bridging the Gap ... 42
Is Real-World Experience Unavoidable? ... 49
Pretraining ... 50
Reinforcement Learning ... 51
Self-Supervised Learning ... 53
Summary ... 54
References ... 55
Chapter 3: Introduction to GANs ... 61
GANs ... 61
CTGAN ... 63
SurfelGAN ... 64
Cycle GANs ... 65
SinGAN-Seg ... 66
MedGAN ... 66
DCGAN ... 67
WGAN ... 68
SeqGAN ... 69
Conditional GAN ... 70
BigGAN ... 71
Summary ... 72
References ... 72

Chapter 4: Synthetic Data Generation with R 75
Basic Functions Used in Generating Synthetic Data 75
Creating a Value Vector from a Known Univariate Distribution 77
Vector Generation from a Multi-Levels Categorical Variable 78
Multivariate 78
Multivariate (with correlation) 79
Generating an Artificial Neural Network Using Package "nnet" in R 84
Augmented Data 90
Image Augmentation Using Torch Package 97
Multivariate Imputation Via "mice" Package in R 102
Generating Synthetic Data with the "conjurer" Package in R 114
Creat a Customer 115
Creat a Product 117
Creating Transactions 118
Generating Synthetic Data 119
Generating Synthetic Data with "Synthpop" Package In R 121
Copula 145
t Copula 147
Normal Copula 150
Gaussian Copula 153
Summary 157
References 157

Chapter 5: Synthetic Data Generation with Python 159
Data Generation with Know Distribution 159
Data with Date information 163
Data with Internet information 163
A more complex and comprehensive example 163
Synthetic Data Generation in Regression Problem 164
Gaussian Noise Apply to Regression Model 168

Friedman Functions and Symbolic Regression 172
Make 3d Plot 174
Make3d Plot 177
Synthetic data generation for Classification and Clustering Problems 182
Classification Problems 183
Clustering Problems 194
Generation Tabular Synthetic Data by Applying GANs 203
Synthetic data Generation 205
Summary 214
Reference 214
Index 215

About the Authors

Necmi Gürsakal a statistics professor at Mudanya University in Turkey, where he shares his experience and knowledge with his students. Before that, he worked as a faculty member at the Econometrics Department Bursa Uludağ University for more than 40 years. Necmi has many published Turkish books and English and Turkish articles on data science, machine learning, artificial intelligence, social network analysis, and big data. In addition, he has served as a consultant to various business organizations.

Sadullah Çelik a mathematician, statistician, and data scientist who completed his undergraduate and graduate education in mathematics and his doctorate in statistics. He has written Turkish and English numerous articles on big data, data science, machine learning, multivariate statistics, and network science. He developed his programming and machine learning knowledge while writing his doctoral thesis, Big Data and Its Applications in Statistics. He has been working as a Research Assistant at Adnan Menderes University Aydin, for more than 8 years and has extensive knowledge and experience in big data, data science, machine learning, and statistics, which he passes on to his students.

Esma Birişçi a programmer, statistician, and operations researcher with more than 15 years of experience in computer program development and five years in teaching students. She developed her programming ability while studying for her bachelor degree, and knowledge of machine learning during her master degree program. She completed her thesis about data augmentation and supervised learning. Esma transferred to Industrial Engineering and completed her doctorate program on dynamic and stochastic nonlinear programming. She studied large-scale optimization and life cycle assessment, and developed a large-scale food supply chain system application using Python. She is currently working at Bursa Uludag University, Turkey, where she transfers her knowledge to students. In this book, she is proud to be able to explain Python's powerful structure.

About the Technical Reviewer

Fatih Gökmenoğlu is a researcher focused on synthetic data, computational intelligence, domain adaptation, and active learning. He also likes reporting on the results of his research.

His knowledge closely aligns with computer vision, especially with deepfake technology. He studies both the technology itself and ways of countering it.

When he's not on the computer, you'll likely find him spending time with his little daughter, whose development has many inspirations for his work on machine learning.

Preface

In 2017, *The Economist* wrote, "The world's most valuable resource is no longer oil, but data," and this becomes truer with every passing day. The gathering and analysis of massive amounts of data drive the business world, public administration, and science, giving leaders the information they need to make accurate, strategically-sound decisions. Although some worry about the implications of this new "data economy," it is clear that data is here to stay. Those who can harness the power of data will be in a good position to shape the future.

To use data ever more efficiently, machine and deep learning—forms of artificial intelligence (AI)—continue to evolve. And every new development in how data and AI are used impacts innumerable areas of everyday life. In other words, from banking to healthcare to scientific research to sports and entertainment, data has become everything. But, for privacy reasons, it is not always possible to find sufficient data.

As the lines between the real and virtual worlds continue to blur, data scientists have begun to generate synthetic data, with or without real data, to understand, control, and regulate decision-making in the real world. Instead of focusing on how to overcome barriers to data, data professionals have the option of either transforming existing data for their specific use or producing it synthetically. We have written this book to explore the importance and meaning of these two avenues through real-world examples. If you work with or are interested in data science, statistics, machine learning, deep learning, or AI, this book is for you.

While deep learning models' huge data needs are a bottleneck for such applications, synthetic data has allowed these models to be, in a sense, self-fueled. Synthetic data is still an emerging topic, from healthcare to retail, manufacturing to autonomous driving. It should be noted that since labeling processes start with real data. Real data, augmented data, and synthetic data all take place in these deep learning processes.

This book includes examples of Python and R applications for synthetic data production. We hope that it proves to be as comprehensive as you need it to be.

—Necmi Gürsakal
— Sadullah Çelik
— Esma Birişçi

Introduction

"The claim is that nature itself operates in a way that is analogous to a priori reasoning. The way nature operates is, of course, via causation: the processes we see unfolding around us are causal processes, with earlier stages linked to later ones by causal relations" [1]. Data is extremely important in the operation of causal relationships and can be described as the "sine qua non" of these processes. In addition, data quality is related to quantity and diversity, especially in the AI framework.

Data is the key to understanding causal relationships. Without data, it would be impossible to understand how the world works. The philosopher David Hume understood this better than anyone. According to Hume, our knowledge of the world comes from our experiences. Experiences produce data, which can be stored on a computer or in the cloud. Based on this data, we can make predictions about what will happen in the future. These predictions allow us to test our hypotheses and theories. If our predictions are correct, we can have confidence in our ideas. If they are wrong, we need to rethink our hypotheses and theories. This cycle of testing and refinement is how we make progress in science and life. This is how we make progress as scientists and as human beings.

Many of today's technology giants, such as Amazon, Facebook, and Google, have made data-driven decision-making the core of their business models. They have done this by harnessing the power of big data and AI to make decisions that would otherwise be impossible. In many ways, these companies are following in the footsteps of Hume, using data to better understand the world around them.

As technology advances, how we collect and store data also changes. In the past, data was collected through experiments and observations made by scientists. However, with the advent of computers and the internet, data can now be collected automatically and stored in a central location. This has led to a change in the way we think about knowledge. Instead of knowledge being stored in our minds, it is now something that is stored in computers and accessed through algorithms.

This change in the way we think about knowledge has had a profound impact on the way we live and work. In the past, we would have to rely on our memory and experience to make decisions. However, now we can use data to make more informed decisions.

For example, we can use data about the past behavior of consumers to predict what they might buy in the future. This has led to a more efficient and effective way of doing business.

In the age of big data, it is more important than ever to have high-quality data to make accurate predictions. However, it is not only the quantity and quality of the data that is important but also the diversity. The diversity of data sources is important to avoid bias and to get a more accurate picture of the world. This is because different data sources can provide different perspectives on the same issue, which can help to avoid bias. Furthermore, more data sources can provide a more complete picture of what is happening in the world.

Machine Learning

In recent years, a method has been developed to teach machines to see, read, and hear via data input. The point of origin for this is what we think of in the brain as producing output bypassing inputs through a large network of neurons. In this framework, we are trying to give machines the ability to learn by modeling artificial neural networks. Although some authors suggest that the brain does not work that way, this is the path followed today.

Many machines learning projects in new application areas began with the labeling of data by humans to initiate machine training. These projects were categorized under the title of *supervised learning*. This labeling task is similar to the structured content analysis applied in social sciences and humanities. Supervised learning is a type of machine learning that is based on providing the machine with training data that is already labeled. This allows the machine to learn and generalize from the data to make predictions about new data. Supervised learning is a powerful tool for many machine learning applications.

The quality of data used in machine learning studies is crucial for the accuracy of the findings. A study by Geiger et al. (2020) showed that the data used to train a machine learning model for credit scoring was of poor quality, which led to an unfair and inaccurate model. The study highlights the importance of data quality in machine learning research. Data quality is essential for accurate results. Furthermore, the study showed how data labeling impacts data quality. About half of the papers using original human annotation overlap with other papers to some extent, and about 70% of the papers that use multiple overlaps report metrics of inter-annotator agreement [2]. This suggests that the data used in these studies is unreliable and that further research is needed to improve data quality.

As more business decisions are informed by data analysis, more companies are built on data. However, data quality remains a problem. Unfortunately, "garbage in, garbage out," which was a frequently used motto about computers in the past, is valid in the sense of data sampling, which is also used in the framework of machine learning. According to the AI logic most employed today, if qualified college graduates have been successful in obtaining doctorates in the past, they will remain doing so in the future. In this context, naturally, the way to get a good result in machine learning is to include "black swans" in our training data, and this is also a problem with our datasets.

A "black swan" is a term used to describe outliers in datasets. It is a rare event that is difficult to predict and has a major impact on a system. In machine learning, a black swan event is not represented in the training data but could significantly impact the results of the machine learning algorithm. Black swans train models to be more robust to unexpected inputs. It is important to include them in training datasets to avoid biased results.

Over time, technological development has moved it into the framework of human decision-making with data and into the decision-making framework of machines. Now, machines evaluate big data and make decisions with algorithms written by humans. For example, a driverless car can navigate toward the desired destination by constantly collecting data on stationary and moving objects around it in various ways. Autonomous driving is a very important and constantly developing application area for synthetic data. Autonomous driving systems should be developed at a capability level that can solve complex and varied traffic problems in simulation. The scenarios we mentioned in these simulations are sometimes made by gaming engines such as Unreal and Unity. Creating accurate and useful "synthetic data" with simulations based on real data will be the way companies will prefer real data that cannot be easily found.

Synthetic data is becoming an increasingly important tool for businesses looking to improve their AI initiatives and overcome many of the associated challenges. By creating synthetic data, businesses can shape and form data to their needs and augment and de-bias their datasets. This makes synthetic data an essential part of any AI strategy. DataGen, Mostly, Cvedia, Hazy, AI.Reverie, Omniverse, and Anyverse can be counted among the startups that produce synthetic data. Sample images from synthetic outdoor datasets produced by such companies can be seen in the given source.

In addition to the benefits mentioned, synthetic data can also help businesses train their AI models more effectively and efficiently. Businesses can avoid the need for costly and time-consuming data collection processes by using synthetic data. This can help businesses save money and resources and get their AI initiatives up and running more quickly.

Who Is This Book For?

The book is meant for people who want to learn about synthetic data and its applications. It will prove especially useful for people working in machine learning and computer vision, as synthetic data can be used to train machine learning models that can make more accurate predictions about real-world data.

The book is written for the benefit of data scientists, machine learning engineers, deep learning practitioners, artificial intelligence researchers, data engineers, business analysts, information technology professionals, students, and anyone interested in learning more about synthetic data and its applications.

Book Structure

Synthetic data is not originally collected from real-world sources. It is generated by artificial means, using algorithms or mathematical models, and has many applications in deep learning, particularly in training neural networks. This book, which discusses the structure and application of synthetic data, consists of five chapters.

Chapter 1 covers synthetic data, why it is important, and how it can be used in data science and artificial intelligence applications. This chapter also discusses the accuracy problems associated with synthetic data, the life cycle of data, and the tradeoffs between data collection and privacy. Finally, this chapter describes some applications of synthetic data, including financial services, manufacturing, healthcare, automotive, robotics, security, social media, marketing, natural language processing, and computer vision.

Chapter 2 provides information about different ways of generating synthetic data. It covers how to generate fair synthetic data, as well as how to use video games to create synthetic data. The chapter also discusses the synthetic-to-real domain gap and how to overcome it using domain transfer, domain adaptation, and domain randomization.

Finally, the chapter discusses whether a real-world experience is necessary for training machine learning models and, if not, how to achieve it using pretraining, reinforcement learning, and self-supervised learning.

Chapter 3 explains the content and purpose of a *generative adversarial network*, or GAN, a type of AI used to generate new data, like training data.

Chapter 4 explores synthetic data generation with R.

Chapter 5 covers different methods of synthetic data generation with Python.

Learning Outcomes of the Book

Readers of this book will learn about the various types of synthetic data, how to create them, and their benefits and challenges. They will also learn about its importance in data science and artificial intelligence. Furthermore, readers will come away understanding how to employ automatic data labeling and how GANs can be used to generate synthetic data. Lastly, readers who complete this book will know how to generate synthetic data using the R and Python programming languages.

Source Code

The datasets and source code used in this book can be downloaded from github.com/apress/synthetic-data-deep-learning.

References

[1]. M. Rozemund, "The Nature of the Mind," in The Blackwell Guide to Descartes' Meditations, S. Gaukroger, John Wiley & Sons, 2006.

[2]. R. S. Geiger et al., "Garbage In, Garbage Out?," in Proceedings of the 2020 Conference on Fairness, Accountability, and Transparency, Jan. 2020, pp. 325–336. doi: 10.1145/3351095.3372862.

CHAPTER 1

An Introduction to Synthetic Data

In this chapter, we will explore the concept of data and its importance in today's world. We will discuss the lifecycle of data from collection to storage and how synthetic data can be used to improve accuracy in data science and artificial intelligence (AI) applications. Next, we will explore of synthetic data applications in financial services, manufacturing, healthcare, automotive, robotics, security, social media, and marketing. Finally, we will examine natural language processing, computer vision, understanding of visual scenes, and segmentation problems in terms of synthetic data.

What Synthetic Data is?

Despite 21st-century advances in data collection and analysis, there is still a lack of understanding of how to properly utilize data to minimize the perceived ambiguity or subjectivity of the information it represents. This is because the same meaning can be expressed in a variety of ways, and a single expression can have multiple meanings. As a result, it is difficult to create a comprehensive framework for data interpretation that considers all of the potential nuances and implications of the information. One way to overcome this challenge is to develop standardized methods for data collection and analysis. This will ensure that data is collected consistently and that the results can be compared across different studies and synthetic data can help us do just that.

People generally view synthetic data as being less reliable than data that is obtained by direct measurement. Put simply, Synthetic data is data that is generated by a computer program rather than being collected from real-world sources. While synthetic data is often less reliable than data that is collected directly from the real world, it is still an essential tool for data scientists. This is because synthetic data can be used to test

N. Gürsakal et al., *Synthetic Data for Deep Learning*, https://doi.org/10.1007/978-1-4842-8587-9_1

hypotheses and models before they are applied to real-world data. This can help data scientists avoid making errors that could have negative consequences in the real world.

Synthetic data that is artificially generated by a computer program or simulation, rather than being collected from real-world sources [11]. When we examine this definition, we see that the following concepts are included in the definition: "Annotated information, computer simulations, algorithm, and "not measured in a real-world".

The key features of synthetic data are as follows:

- Not obtained by direct measurement
- Generated via an algorithm
- Associated with a mathematical or statistical model
- Mimics real data

Now let's explain why synthetic data is important.

Why is Synthetic Data Important?

Humans have a habit of creating synthetic versions of expensive products. Silk is an expensive product that began to be used thousands of years ago, and rayon was created in the 1880s. So, it's little surprise that people would do the same with data, choosing to produce synthetic data because it is cost-effective. As mentioned earlier, synthetic data allows scientists to test hypotheses and models in a controlled environment it can also be used to create "what if" scenarios, helping data scientists to better understand the outcomes of their models.

Likewise, synthetic data can be used in a variety of ways to improve machine learning models and protect the privacy of real data. First, synthetic data can be used to train machine learning models when real data is not available. This is especially important for developing countries, where data is often scarce. Second, synthetic data can be used to test machine learning models before deploying them on real data. This is important for ensuring that models work as intended and won't compromise the real data. Finally, synthetic data can be used to protect the privacy of real data by generating data that is similar to real data but does not contain any personal information.

Synthetic data also provides more control than real data. Actual data comes from many different sources, which can result in datasets so large and diverse that they become unwieldy. Because synthetic data is created using a model whose data is generated for a specific purpose, it will not be randomly scattered. In some cases, synthetic data may even be of a higher quality than real data. Actual data may need to be over-processed when necessary, and too much data may be processed when necessary. These actions can reduce the quality of the data. Synthetic data, on the other hand, can be of higher quality, thanks to the model used to generate the data.

Overall, synthetic data has many advantages over real data. Synthetic data is more controlled, of higher quality, and can be generated in the desired quantities. These factors make synthetic data a valuable tool for many applications. A final reason why synthetic data is important is that it can be used to generate data for research purposes, allowing researchers to study data that is not biased or otherwise not representative of the real data.

Now let's explain the importance of synthetic data for data science and artificial intelligence.

Synthetic Data for Data Science and Artificial Intelligence

The use of synthetic data is not a new concept. In the early days of data science and AI, synthetic data was used to train machine learning models. However, synthetic data of the past was often low-quality and not realistic enough to be useful for training today's more sophisticated AI models.

Recent advances in data generation techniques, such as *Generative Adversarial Networks* (GANs), have made it possible to generate synthetic data that is virtually indistinguishable from real-world data. This high-quality synthetic data is often referred to as "*realistic synthetic data*".

The use of realistic synthetic data has the potential to transform the data science and AI fields. Realistic synthetic data can be used to train machine learning models without the need for real-world data. This is especially beneficial in cases where real-world data is scarce, expensive, or difficult to obtain.

In addition, realistic synthetic data can be used to create "virtual environments" for testing and experimentation. These virtual environments can be used to test machine learning models in a safe and controlled manner, without the need for real-world data.

For example, a computer algorithm might be used to generate realistic-looking images of people or objects. This could be used to train a machine learning system to better recognize these objects in real-world images. Alternatively, synthetic data could be used instead of real-world data if the latter is not available or is too expensive to obtain.

Overall, the use of synthetic data is a promising new trend in data science and AI. The ability to generate high-quality synthetic data is opening new possibilities for training and experimentation. For example, a synthetic data set could be created that contains 6000 words, instead of the usual 2000. This would allow the AI system to learn from a larger and more diverse data set, which would in turn improve its performance on real-world data. In the future, synthetic data is likely to play an increasingly important role in the data science and AI fields.

Let us now consider accuracy problems in terms of synthetic data.

Accuracy Problems

Supervised learning algorithms are trained with labeled data. In this method, the data is commonly named "*ground truth*", and the test data is called "*holdout data*". We have three types to compare accuracies across algorithms [2]:

- Estimator score method: An estimator is a number that is used to estimate, or guess, the value of something. The score method is a way to decide how good an estimator is by analyzing how close the estimator's guesses are to the actual value.
- Scoring parameter: Cross-validation is a model-evaluation technique that relies on an internal scoring strategy.
- Metric functions: The sklearn.metrics module provides functions for assessing prediction error for specific purposes.

It's important to acknowledge that the accuracy of synthetic data can be problematic for several reasons. First, the data may be generated by a process that is not representative of the real-world process that the data is meant to represent. This can lead to inaccuracies in the synthetic data that may not be present in real-world data. Second, the data may be generated with a specific goal in mind, such as training a machine learning algorithm, that does not match the goal of the data's user of the synthetic data. This can also lead to inaccuracies in the synthetic data. Finally, synthetic data may be generated using a stochastic process, which can introduce randomness into the data that may not be present in real-world data. This randomness can also lead to inaccuracies in the synthetic data.

One way to overcome potential issues with accuracy in machine learning is to use synthetic data. This can be done by automatically tagging and preparing data for machine learning algorithms, which cuts down on the time and resources needed to create a training dataset. This also creates a more consistent dataset that is less likely to contain errors. Another way to improve accuracy in machine learning is to use a larger training dataset. This will typically result in better performance from the machine learning algorithm.

Working on the recognition and classification of aircraft from satellite photos, the Airbus and OneView companies, in their studies on data for machine learning, achieved accuracy of 88% versus 82% with the simulated dataset of OneView company, compared to data consisting of only of real data. When real data and synthetic data are used in a mixed way, an accuracy of ~ 90% is obtained, and this number represents an 8% improvement over real-only real data [3]. This improved accuracy is due to the increased variety of data that is available when both real and simulated data are used. The increased variety of data allows the machine learning algorithm to better learn the underlying patterns of the data. This improved accuracy is significant and can lead to better decision-making in a variety of applications.

Now let's examine the life cycle of data in terms of synthetic data.

The Lifecycle of Data

Before leveraging the power of synthetic data, it's important to understand the lifecycle of data. First, it can help organizations to better manage their data; by understanding the stages that data goes through, organizations can more effectively control how data is used and prevent unauthorized access. Additionally, the data lifecycle can help organizations ensure that their data is of high quality. Finally, the data lifecycle can help organizations plan for the eventual destruction of data; by understanding when data is no longer needed, organizations can ensure that they do not keep data longer than necessary, which can both save space and reduce costs.

The data lifecycle is the process of managing data from its creation to its eventual disposal. Figure 1-1 shows the five main phases of the data lifecycle.

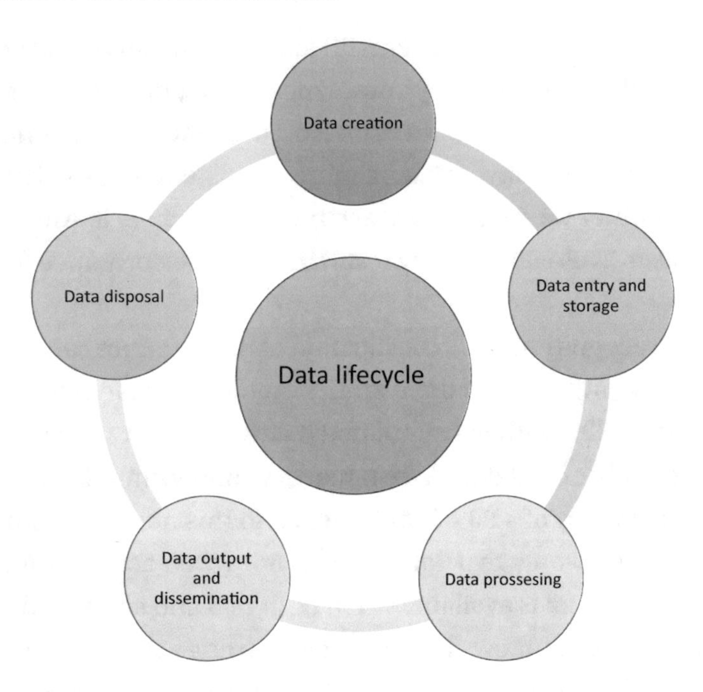

Figure 1-1. *Data lifecycle*

In the context of synthetic data, the data life cycle refers to the process of generating, storing, manipulating, and outputting synthetic data. This process is typically carried out by computers and involves the use of algorithms and rules to generate data that resembles real-world data.

Following are the five stages of the data lifecycle:

1. **Data creation:** The first stage of the data lifecycle is data creation. This is the stage at which synthetic data is first generated, either through direct input or through the capture of information from an external source.

2. **Data entry and storage:** This stage involves the entry of synthetic data into a computer system and its storage in a database. Data entry and storage typically involve the use of algorithms or rules to generate data that resembles real-world data.

3. **Data processing:** This stage covers the manipulation of synthetic data within the computer system, to convert it into a format that is more useful format for users. This may involve the use of algorithms and the application of rules and filters. Data processing typically involves the use of algorithms or rules to generate data that resembles real-world data.

4. **Data output and dissemination:** This stage is the process of generating synthetic data from a computer system and making it available to users. This may involve the generation of reports, the creation of graphs and charts, or the output of data in a format that can be imported into another system.

5. **Data disposal:** The final stage of the data lifecycle is data disposal. This stage covers the disposal of synthetic data that is no longer needed. This may involve the deletion of data from a database or the physical destruction of storage media. Data disposal typically involves the use of algorithms or rules to generate.

Reinforcement learning algorithms are used to learn how to do things by interacting with an environment. However, these algorithms can be inefficient, meaning they need a lot of interactions to learn well. To address this issue, some people are using external sources of knowledge, such as data from demonstrations or observations. This data can come from experts, real-world demonstrations, simulations, or synthetic demonstrations.

Researchers at Google and DeepMind think that datasets also have a lifecycle and they summarize the lifecycle of datasets in three stages: Producing the data, consuming the data, and sharing the data [4].

In the production data phase, users record their interactions with the environment and provide datasets. At this stage, users add additional information to the data automatically or manually labeling or filtering the data.

In the consuming the data phase, researchers analyze and visualize datasets or use them to train algorithms for machine learning purposes.

In the sharing the data stage, researchers often share their data with other researchers to help with their reserach. When researchers share data, it makes it easier for other researchers to run and validate new algorithms. However, the researchers who produced the data still own it and should be given credit for their work.

Let's now consider data collection and privacy issues in terms of synthetic data.

Data Collection versus Privacy

Data can be collected in many ways. For example, data from radar, LIDAR, and the camera systems of driverless cars can be taken and fused into a format usable for decision making. Considering that fusion data is also virtual data, it is necessary to think in detail about the importance of real and virtual data. So, since real data can be converted into virtual data, and used or augmented data can be used together with real data in machine learning, these two data types are important for us. This is sometimes referred to as the **Ensemble Method**.

In the Ensemble Method, a few basic models can be combined to create an optimal predictive model; Why can't the data be fused, and more qualified data can be obtained?

Labeling the data is tedious and costly, as machine learning models require large and diverse data to produce good results. Therefore creating synthetic data by transforming real data using with data augmentation techniques or directly generating synthetic data and using it as an alternative to real data can reduce transaction costs. According to Gartner, by 2030, there will be more synthetic data than real data in AI models.

Another issue that synthetic data can help overcome is that of data privacy.

Data Privacy and Synthetic Data

Today, many institutions and organizations use large amounts of data to forecast, create policies, plan, and achieve higher profit margins. By using this data, they can better understand the world around them and make more informed decisions. However, due to privacy restrictions and guarantees given to personal data, only the personnel of the institutions have full access to such data. Anonymization techniques are used to prevent the identities of data subjects from being revealed. Sure, data collectors can maintain data privacy by using aggregation, recoding, record exchange, suppression of sensitive values, and random error insertion, data collectors can maintain data privacy. However, advances in computer and cloud technologies are likely to make such measures insufficient to maintain data privacy. We'll explore some examples in the next section.

In today's world, with the advances in information technology, patient data, and driver the data, of those using vehicles the data obtained by research companies from public opinion surveys have reached enormous amounts. However, most of the time, when this data is used to find new solutions, the concept of "individual privacy" comes up. This problem is overcome by anonymizing the data, which is the process of modifying the data to eliminate any information that could lead to privacy intrusion. Anonymizing data is important to protect people's privacy, as even without personal identifiers, the remaining attributes in the data may still be used to re-identify an individual. In the simplest form of data anonymization, all personal identifiers are removed. However, it has been shown that this is not enough to protect people's privacy. Therefore, it is important to fully anonymize data possible to protect people's privacy rights.

Most people think that privacy is protected by anonymization. This is when there is no name, surname, or any sign to indicate identity in the database. However, this is not always accurate. This means that if you have an account on both Twitter and Flickr, there's a good chance that someone could identify you from the anonymous Twitter graph. However, the error rate is only 12%, so the chances are still pretty good that you won't be identified [14]. Even though the chances of being identified are relatively low, it is still important to be aware of the potential risks of sharing personal information online. Anonymity is not a guarantee of privacy, and even seemingly innocuous information can identify individuals in certain cases. Therefore, exercising caution is required when sharing personal information online, even if it is ostensibly anonymous.

Anonymization and labeling are two primary techniques in AI applications. However, both techniques have their own set of problems. Anonymization can lead to the loss of vital information, while labeling can introduce bias and be costly to implement. In addition, hand-labeled data might not be high quality because it is often mislabeled. To overcome these problems, researchers have proposed various methods, such as semi-supervised learning and active learning. However, these methods are still not perfect, and further research is needed to improve them.

The Bottom Line

The collection of more data with increasing data sources makes it necessary for businesses to take security measures against information attacks. In some cases, businesses need more data than is available to innovate in certain areas. In some cases, more data may be necessary due to a lack of practical research or high costs of data collection. Many businesses generate data programmatically in the real world to obtain otherwise unattainable information. The use of synthetic data is becoming increasingly popular as businesses attempt to collect more data and test different scenarios. Synthetic data is created by computer programs and is designed to mimic real-world data. This allows businesses to gather data more efficiently and to test various scenarios to see what may happen in the real world.

The world is becoming more data-centric, so businesses are starting to use computer programs to create data similar to data gathered from the real world. This is useful because it facilitates data collection and helps businesses test different scenarios to see what will happen in the real world.

Now let's examine synthetic data and data quality.

Synthetic Data and Data Quality

When working on AI projects, it is important to focus on data quality. It all starts there; if data quality is poor, the AI system will be fatally compromised. Data cascades can occur when AI practitioners apply conventional AI practices that don't value data quality. Most AI practitioners (92%) have reported experiencing one or more data cascades. This often happens because they applied conventional AI practices that didn't value data quality. For this reason, it is important to use high-quality data when training deep learning networks [5]. Andrew Ng has said that "Data is food for AI" and that the issue of data quality should be focused on data more than the modei/algorithm [6]. The use of synthetic data can help to address the issue of data quality in AI projects. This is because synthetic data can be generated to be of high quality, and it can be generated to be representative of the real-world data that the AI system will be used on. This means that the AI system trained on synthetic data will be more likely to generalize well to the real world.

AI technologies in particular use synthetic data intensively. Just a few examples include medicine, where synthetic data is used extensively to test specific conditions and cases for which real data is not available; self-driving cars, such as the ones used by Uber and Google, are trained using synthetic data; fraud detection and protection in the financial industry is facilitated using synthetic data. Synthetic data gives data professionals access to centrally stored data while maintaining the privacy of the data. In addition, synthetic data reproduce important features of real data without revealing its true meaning and protects confidentiality. In research departments, on the other hand, synthetic data is used to develop and deliver innovative products for which the necessary data may not be available [7]. Overall, the use of synthetic data is extremely beneficial as it allows for the testing of new products and services while maintaining the privacy of the original data. Synthetic data is also incredibly versatile and can be used in a variety of different industries and applications. In the future, the use of synthetic data is likely to become even more widespread as the benefits of using it become more widely known.

Let us now examine some of synthetic data applications.

Aplications of Synthetic Data

Synthetic data is often used in financial services, manufacturing, healthcare, automotive, robotics, security, social media, and marketing.

Let's first quickly explore how synthetic data cen be used in finance.

Financial Services

The use of synthetic data is becoming increasingly important in financial services as the industry moves towards more data-driven decision-making. Synthetic data can be used to supplement or replace traditional data sources, providing a more complete picture of the underlying risk.

Financial services is an industry that is highly regulated and subject to constant change. New rules and regulations are constantly being introduced, and the industry is constantly evolving. As a result, it can be difficult for financial institutions to keep up with the changes and ensure that their data is compliant.

Synthetic data can be used to generate data that is compliant with the latest rules and regulations. This can help financial institutions avoid the costly fines and penalties that can be associated with non-compliance. In addition, synthetic data can be used to test new products and services before they are launched. This can help financial institutions avoid the costly mistakes that can be made when launching new products and services.

Synthetic data can also help to improve the accuracy of risk models by providing a more complete picture of underlying risks. For example, consider a portfolio of loans. Traditional data sources may only provide information on the loan amount, interest rate, and term. However, synthetic data can provide additional information on the borrower's credit score, employment history, and other factors that can impact the risk of default. This additional information can help to improve the accuracy of the risk model.

Another key benefit of synthetic data is that it can provide a way to test and validate models before they are deployed in live environments. This is because synthetic data can be generated with known values for the inputs and outputs. This allows for the testing of models under a variety of different scenarios, which can help to identify any potential issues before the model is deployed in a live environment.

Synthetic data can be used in financial services in a variety of other ways. For example, it can be used to:

- **Generate realistic scenarios for stress testing and risk management:** Generating synthetic data can help financial institutions to identify potential risks and to develop plans for dealing with them. This can be used to generate realistic scenarios for stress testing and risk management purposes. Doing so can help to improve the resilience of the financial system.

- **Train machine learning models:** Synthetic data can help train machine learning models for tasks such as fraud detection and credit scoring. This can automate processes for financial institutions and make them more efficient.
- **Generate synthetic transactions:** Synthetic data can be used to generate synthetic transactions, which can help financial institutions test new products and services, or simulate market conditions.
- **Generate synthetic customer data:** Financial institutions can use synthetic data to generate synthetic customer data. This can help them to test new customer acquisition strategies or to evaluate customer service levels.
- **Generate synthetic financial data:** Synthetic data can be used to generate synthetic financial data. This can help financial institutions to test new financial products or to evaluate the impact of new regulations.

Finally, synthetic data can help to reduce the cost of data acquisition and storage. This is because synthetic data can be generated on-demand, as needed. This eliminates the need to store large amounts of data, which can save on both the cost of data acquisition and storage.

Now, let's look at how synthetic data can be used in the manufacturing field.

Manufacturing

In the world of manufacturing, data is used to help inform decision-makers about various aspects of the manufacturing process, from production line efficiency to quality control. In some cases, this data is easy to come by- for example, data on production line outputs can be gathered through sensors and other monitoring devices. However, in other cases, data can be much more difficult to obtain. For example, data on the performance of individual components within a production line may be hard to come by or may be prohibitively expensive to gather. In these cases, synthetic data can be used to fill in the gaps.

In many manufacturing settings, it is difficult or impossible to obtain real-world data that can be used to train models. This is often due to the proprietary nature of manufacturing processes, which can make it difficult to obtain data from inside a factory. Additionally, the data collected in a manufacturing setting may be too noisy or unrepresentative to be useful for training models.

To address these issues, synthetic data can be used to train models for manufacturing applications. However, it is important to consider both the advantages and disadvantages of using synthetic data before deciding whether it is the right choice for a particular application.

Synthetic data can be employed in manufacturing in several ways. First, synthetic data can be used to train machine learning models that can be used to automate various tasks in the manufacturing process. This can improve the efficiency of the manufacturing process and help to reduce costs. Second, synthetic data can be used to test and validate manufacturing processes and equipment. This can help to ensure that the manufacturing process is running smoothly, and that the equipment is operating correctly. Third, synthetic data can be used to monitor the manufacturing process and to identify potential problems. This can help to improve the quality of the products being produced and to avoid costly manufacturing defects.

Synthetic data can be used to improve the efficiency of data-driven models. This is because synthetic data can be generated much faster than real-world data. This is important because it is allowing manufacturers to train data-driven models faster and get them to market quicker.

The use of synthetic data is widespread in the manufacturing industry. It helps companies to improve product quality, reduce manufacturing costs, and improve process efficiency. Some examples of the use of synthetic data in manufacturing are as follows:

- **Quality Control**: Synthetic data can be used to create models that predict the likelihood of defects in products. This information can be used to improve quality control procedures.
- **Cost Reduction**: The use of synthetic data can help identify patterns in manufacturing processes that lead to increased costs. This information can be used to develop strategies for reducing costs, thereby reducing the overall cost of production.
- **Efficiency Improvement**: Synthetic data can be used to create models that predict the efficiency of manufacturing processes. This information can be used to improve process efficiency.

- **Product Development**: Synthetic data can help improve product development processes by predicting the performance of new products. In this way, it can be decided which products to monitor and how to develop them.
- **Production Planning**: Production planning can be done by using synthetic data to create models that predict the demand for products. In this way, businesses can improve their production planning by making better predictions about future demand.
- **Maintenance**: Synthetic data can be used to create models that predict the probability of equipment failures. In this way, preventive measures can be taken, and maintenance processes can be improved by predicting when equipment will fail.

Now, let's quickly explore how synthetic data can be employed in the healthcare realm.

Healthcare

The most obvious benefit of utilizing synthetic data in healthcare is to protect the privacy of patients. By using synthetic data, healthcare organizations can create models and simulations that are based on real data but do not contain any actual patient information. This can be extremely helpful in situations where patient privacy is of paramount concern, such as when developing new treatments or testing new medical devices.

The use of synthetic data will evolve in line with the needs and requirements of health institutions. However, the following are some of the most common reasons why healthcare organizations might use synthetic data include:

- Machine learning models: One of the most common reasons why healthcare organizations use synthetic data is to train machine learning models. This is because synthetic data can be generated in a controlled environment, which allows for more reliable results.
- Artificial intelligence: synthetic data can be used to identify patterns in patient data that may be indicative of a particular condition or disease. This can then be used to help diagnose patients more accurately and to also help predict how they are likely to respond

to treatment. This is extremely important in terms of ensuring that patients receive the most effective care possible.

- Protect privacy: One of the biggest challenges in the healthcare industry is the reliable sharing of data. Health data is very important for doctors to diagnose and treat patients quickly. For this reason, many hospitals and health institutions attach great importance to patient data. Synthetic data help provide the best possible treatment. In addition, synthetic data is a technology that will help healthcare organizations share information while protecting personal privacy.
- Treatments: Another common reason why healthcare organizations use synthetic data is to test new treatments. This is because synthetic data can be used to create realistic simulations of real-world conditions, which can help to identify potential side effects or issues with a new treatment before it is used on real patients.
- To help design new drugs and to test their efficacy.
- Improve patient care: Healthcare organizations can also use synthetic data to improve patient care. This is because synthetic data can be used to create realistic simulations of real-world conditions, which can help healthcare professionals to identify potential issues and make better-informed decisions about patient care.
- Reduce costs: Healthcare organizations can also use synthetic data to reduce cost. This is because synthetic data can be generated relatively cheaply, which can help to reduce the overall costs associated with real-world data collection and analysis.
- Several hospitals are now using synthetic data in the health sector to improve the quality of care that they can provide. This is being one in several different ways, but one of the most common is to use computer simulations. This allows for a more realistic representation of patients and their conditions, which can then be used to test out new treatments or procedures. This can be extremely beneficial in reducing the risk of complications and ensuring that patients receive the best possible care.

Overall, the use of synthetic data in the health sector is extremely beneficial. It is helping to improve the quality of care that is being provided and is also helping to reduce the risk of complications. In addition, it is also helping to speed up the process of diagnosis and treatment.

Now let's look at how synthetic data can be used in the automotive industry field.

Automotive

Another application of synthetic data in the automotive industry is autonomous driving. A large amount of data is needed to train an autonomous driving system. This data can be used to train a machine learning model that can then be used to make predictions about how the autonomous driving system should behave in different situations. However, real-world data is often scarce, expensive, and difficult to obtain.

Another important application of synthetic data in automotive is in safety-critical systems. To ensure the safety of a vehicle, it is important essential to be able to test the systems in a variety of scenarios. Synthetic data can be used to generated data for all the different scenarios that need to be tested. This is important because it allows for to provide more thorough testing of system and helps ensure the safety of the vehicle.

Overall, synthetic data has to potential to be a valuable tool for the automotive industry. It can be used to speed up the development process and to generate large quantities of data. However, it is important to be aware of the challenges associated with synthetic data and to ensure that it is used in a way that maximizes its benefits.

There are a few reasons why automotive companies need synthetic data. The first has to do with the development of new technologies a large amount of data. In order to create and test new features or technologies, companies need a large amount of data. This data is used to train algorithms that will eventually be used in the product. However, collecting this data can be difficult, time-consuming, and expensive.

Another reason automotive companies need synthetic data is for testing purposes. Before a new product is released, it needs to go through rigorous testing. This testing often includes putting the product through a range of different scenarios. However, it can be difficult to test every single scenario in the real world. This is where synthetic data comes in. It can be used to create realistic test scenarios that would be difficult or impossible to re-create in the real world.

Synthetic data can be used for marketing purposes. Automotive companies also often use data to create marketing materials such as ads or website content. However, this data can be difficult to obtain. Synthetic data can be used to create realistic marketing scenarios that can be used to test different marketing strategies.

In conclusion, synthetic data is needed in automotive industry for a variety of reasons. It can be used to create realistic test scenarios, train algorithms, and create marketing materials.

Now let's look at how synthetic data is used in the robotics field.

Robotics

Robots are machines that can be programmed to do specific tasks. Sometimes these tasks are very simple, like moving a piece of paper from one place to another. Other times, the tasks are more complex, like moving around in the world and doing things that humans can do, like solving a Rubik's Cube. Creating robots that can do complex tasks is a challenge because the robots need a lot of training data to behave like humans. This data can be generated by simulations, which is a way of creating a model of how the robot will behave.

There are several reasons why synthetic data is needed in robotics. The first is that real-world data is often scarce. This is especially true for data needed to train machine learning models, a key component of robotics. Synthetic data can be used to supplement real-world data and, in some cases, to replace them entirely. Second, real-world data is often noisy. This noise can come from a variety of sources, such as sensors, actuators, and the environment. Synthetic data can be used to generate noise-free data that can be helpful for training machine learning models. The third reason is that collecting real-world data is often expensive. This is especially true for data needed to train machine learning models. Synthetic data can be used to generate data that is much cheaper to collect. A fourth reason is that real-world data is often biased. This bias can come from a variety of sources, such as sensors, actuators, and the environment. Synthetic data can be used to generate bias-free data that can be helpful for training machine learning models. The fifth reason synthetic data is needed in robotics is that real-world data is often unrepresentative. This is especially true for data needed to train machine learning models. Synthetic data can be used to create data that better represents the real world, which can be helpful for training machine learning models.

Robots can learn to identify and respond to different types of objects by using synthetic data. By learning from this data, the robot can learn how to better identify and respond to different types of human behavior. For example, a robot might be given a set of synthetic data that includes variousa variety of different types of human behavior and how to respond to them.

Now let's look at how synthetic data can be used in security field.

Security

Synthetic data can play a vital role in enhancing security, both through its ability to train machine learning models to better detect security threats and by providing it also provides a means way of testing security systems and measuring their effectiveness.

Machine learning models that are trained on synthetic data are more effective at detecting security threats because they are not limited by available the real-world data that is available synthetic data can be generated to match any desired distribution, including distributions that are not present in the real world. This allows machine learning models to learn more about the underlying distribution of data, and to better identify outliers that may represent security threats.

Testing security systems with synthetic data is important because it allows a controlled environmentets for measure the system's performance. Synthetic data can be generated to match any desired distribution of security threats, making it possible to test how well a security system can detect and respond to a wide variety of threats. This is importnat because real-world data is often limited in scope and may not be representative of the full range of security threats that a system may encounter.

Overall, the use of synthetic data is importnat essential for both training machine learning models to detect security threats and for testing the performance of security systems. Synthetic data provides a more complete picture of the underlying distribution of data which leads to better improves the detection of security threats. Additionally, synthetic data can be used to create controlled environments for testing security system performance, making it possible to measure the effectiveness of a security system more accurately.

Now, let's quickly explore how synthetic data can be employed in the social media realm.

Social Media

Social media has become an integral part of our lives. It is a platform where we share our thoughts, ideas, and feelings with our friends and family. However, social media has also become a breeding ground for fake news and misinformation. This is because anyone can create a fake account and spread false information.

To combat this problem, many social media platforms are now using AI to detect fake accounts and flag them. However, AI can only be as effective as the data it is trained on. If the data is biased or inaccurate, the AI will also be biased or inaccurate. This is where synthetic data comes in. Synthetic data can be used to train AI algorithms to be more accurate in detecting fake accounts. Synthetic data can help reduce the spread of fake news and misinformation on social media.

One way to generate synthetic data is to use generative models. For example, a generative model could be trained on a dataset of real images of people. Once trained, the model could then generate new images of people that look real but are fake. This is important because it allows us to creates data that is representative of the real world.

Simulation is another way of generating synthetic data. For example, we could create a simulation of a social media platform. This simulation would include all the same features as the real social media platform. However, it would also allows us to control what data is generated. This is important because it allows us to test different scenarios. For example, we could test what would happen if a certain percentage of accounts were fake. This would allow us to see how our AI algorithms would react in the real world.

Some social media platforms that have been known to use synthetic data include Facebook, Google, and Twitter; Each of this platforms has used synthetic data in different ways and for different purposes.

Facebook has been known to uses synthetic data to train its algorithms. For example, Facebook has used synthetic data to train its facial recognition algorithms. Because it is difficult to obtain a large enough dataset of real-world faces to train these algorithms effectively. In addition, Facebook has also used synthetic data to generate fake user profiles. This is done to test how effective plartfom algorithms are at detecting fake profiles.

In addition to using real data, Google has been known to use synthetic data. Synthetic data is generated data that is designed to mimic real data. For example, Google has to used synthetic data to train its machine learning algorithms to better understand natural language. Google has also used synthetic data to generate fake reviews. This is done to test how effective the platform's algorithms at detectare detecting fake reviews.

Twitter is also known to use synthetic data. The platform has used synthetic data to generate fake tweets and fake user profiles to test how effective its algorithms are at detecting detect them.

Now, let's quickly explore how synthetic data can be employed in the marketing realm.

Marketing

There are many benefits to using synthetic data in marketing. Perhaps the most obvious benefit is that it can be used to generate data that would be otherwise unavailable. This is especially useful for marketing research, as because it can be used to generate data about consumer behavior that would be difficult or impossible to obtain through traditional means.

The use of synthetic data in marketing is important for several reasons. First, it allows marketing researchers to study behavior in a controlled environment. This is important because it allows for the isolation of variables and the testing of hypotheses in a way that would not be possible with real-world data. Second, synthetic data can be used to generate new insights into consumer behavior. By analyzing how consumers behave in a simulated environment, marketing researchers can develop new theories and models that can be applied to real-world data. Finally, synthetic data can be used to evaluate marketing campaigns and strategies. By testing campaigns and strategies in a simulated environment, marketers can identify which ones are most likely to be successful in the real world.

Synthetic data is also needed in marketing because it can be used to protect the privacy of real customers. By using synthetic data instead of real customer data, marketers can avoid having to collect and store sensitive information about their customers. This is especially important for businesses that are subject to strict privacy laws, such as those in the European Union.

Several marketing organizations use synthetic data to get a better understanding of customer behavior and to improve marketing strategies. Each of these organizations uses synthetic data in different ways, but all of them utilize it to gain insights into the behavior of consumers.

Natural Language Processing

Language models are trained on a large corpus of text and can be used to generate new text that is similar to the training data. Language models can be used to generate synthetic data that is representative of different groups of people [8] or to generate data with specific properties.

Natural language processing (NLP) is a subfield of AI that deals with the interpretation and manipulation of human language. NLP is used in a variety of applications, including text classification, chatbots, and machine translation. NLP helps computers to understand, interpret, and manipulate human language [9].

One area where NLP will likely have a significant impact is in the generation of synthetic data. Synethetic data that is generated by artificial means, as opposed to being collected from real-world sources. Synthetic data can be used to train machine learning models [10], and NLP can be used to generate synthetic data that is realistic and diverse. This is important because it allows machine learning models to be trained on data that is representative of the real world, which can improve the accuracy of the models. For example, synthetic data can be used to generate realistic images of people or objects that don't exist in the real world or to create simulated environments for training autonomous vehicles. Machine learning models trained on data that is representative of the real world have improved accuracy.

In addition, NLP can be used to automatically label synthetic data, which is important for training supervised machine learning models. For example, NLP can be used to also automatically generate descriptions of images or videos, which can then be used as labels for training image recognition or object detection models. This is important for training supervised machine learining models, as it can help reduce the amount of manual labeling that is required.

Overall, NLP is a powerful tool for generating and manipulating synthetic data. It can be used to automatically generate large amounts of realistic data, which is important for training machine learning models. In addition, NLP can automatically label synthetic data, which is important for training supervised machine learning models.

Consequently, in the feature, NLP will continue to play an importnat role in the generation of synthetic data. The use of NLP to generate synthetic data will allow for the creation of data that is more representative of different groups of people and will allow for the creation of data with specific properties.

Computer Vision

Computer vision is the process of using computers to interpret and understand digital images. This can be done using algorithms that can identify patterns and features in images, and then make decisions about what those images represent.

The computer can detect a human by identifying their silhouette or by drawing bounding boxes around the person. In the photos in Figure 1-2, the computer would count the person by identifying their silhouette, as seen in Figure 1-2 (a) (instance segmentation) or by drawing bounding boxes around the person (object detection), as in Figure 1-2 (b).

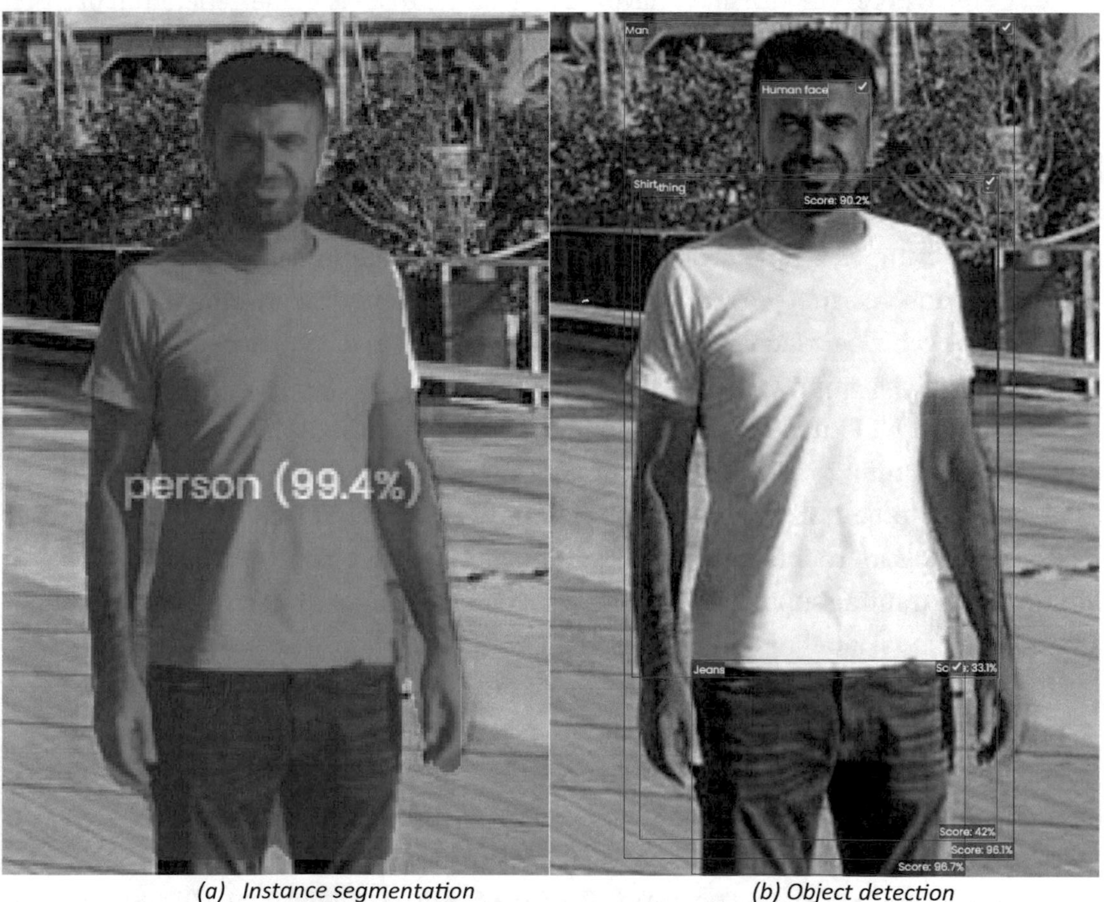

(a) Instance segmentation (b) Object detection

Figure 1-2. *Computer vision: (a) instance segmentation (b) object detection*

An autonomous vehicle needs to be able to detect objects and understand what they are to make decisions in real-time. This is done through object detection, which is the process of finding bounding boxes around each item in the image. After that, semantic segmentation assigns a label to every pixel in the image, indicating what the object is. Finally, instance segmentation shows you how many individual objects there are [11]. Functions such as seeing, reading, and hearing on a computer are performed with artificial neural networks that imitate the human brain. Recognition of human faces, navigation of autonomous cars, and diagnosing patients with the help of medical images obtained by scanning on the computer are all related to the computer's vision, and these works are done with algorithms called *artificial neural networks*.

The neural network is an area type of machine learning algorithm that is used to simulate the workings of the human brain. It is made up of a series of interconnected nodes, each with its weight and three should value. If the output of a node is above the specified threshold, that node is activated and sends data to the next layer of the network. If the output of a node is below the specified threshold, that node is not activated, and no data is sent along [12]. The output of the node is compared to the threshold. If the output is above the threshold, the node is activated and sends data to the next layer.

When a model learns too much detail and noise from the training data, it is said to be overfitting. This can negatively impact the model's performance on new data [13]. This means that if we use synthetic data when training a deep learning model, it will be less likely to overfit the data and as a result, the model will be more accurate and able to generalize to new data better.

Convolutional Neural Networks (CNN) are specifically designed to be good at image recognition tasks. They work by taking in an image as input and then assigning importance to various aspects or objects in the image. This allows the CNN to differentiate between different objects in the image. Such a deep neural network has a much more complex structure than a simple neural network. The following resources can be reviewed for CNN applications in R and Python.

However, no matter how complex the algorithms are in computer vision; as we mention before the quality of the data, which is only called "*input*" in such forms, is very important for the accuracy of the results. Whether the data is text, audio or photographic, the size of the data and its good labeling are essential to the accuracy of the results. The raw material of artificial intelligence is big data. Privacy and data privacy can prevent companies from accessing big data with appropriate laws and enforcement. In addition, labeling data for training purposes also creates a significant cost in machine learning.

Understanding of Visual Scenes

The Visual scenes are the images that are seen on the computer screen. These images are made up of pixels, which are tiny dots of color. The pixels are arranged in a grid, and each pixel has a specific address. The computer visual scenes are created by the computer graphics software, which is responsible for creating the images that are seen on the screen. The computer graphics software can create the images by using the addresses of the pixels.

One of the most important purposes of computer vision is to examine visual scenes and understand them in detail. This includes determining the objects in the image, their places in 2D and 3D, their attributes, and making a semantic description of the scene.

Different types of tasks that can be performed with object recognition datasets: (a) Image classification involves assigning a label to an entire image, (b) object bounding box localization involves identifying the location of an object within an image, and (c) semantic pixel-level segmentation involves classifying each pixel in an image, and (d) segmenting individual object instances. This involves identifying each object in an image and then segmenting it from the rest of the image, as demonstrated in Figure 1-3.

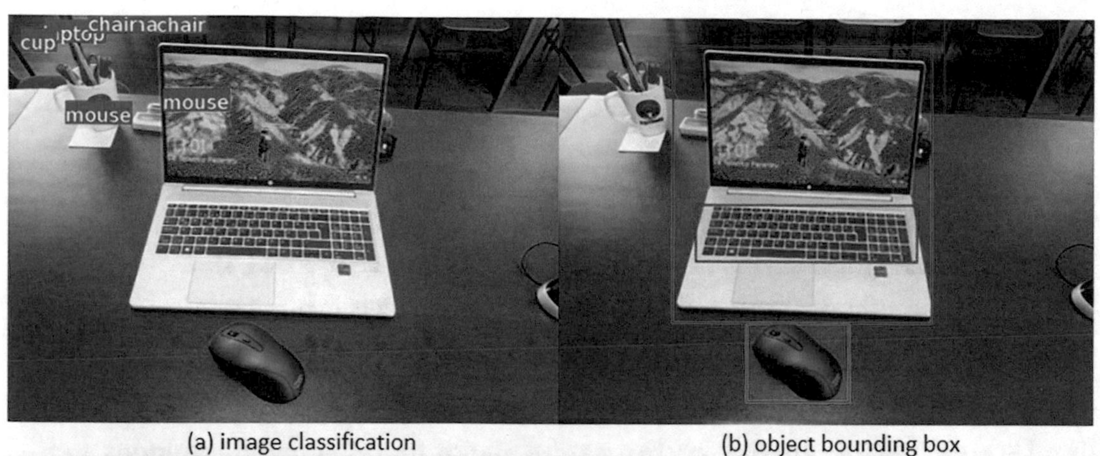

(a) image classification (b) object bounding box

(c) semantic pixel-level segmentation (d) segmenting individual object instances

Figure 1-3. *Object recognition*

We can explain what us done in the four panels shown in of Figure 1-3 as fallows:

(a) **Image Classification:** This task is done by giving binary labels to objects so that they can be recognized as they exist in the image, as shown in Figure 1-3 (a). It is widely used in classification, image processing, and machine learning.

(b) **Object bounding box localization:** It is a method used to determine the boundaries of an object, as shown in Figure 1-3 (b). This method is used to detect where an object in an image begins and ends.

(c) **Semantic pixel-level segmentation:** It is a method used to distinguish objects in an image from each other, as shown in Figure 1-3 (c). This method is used to detect whether each pixel belongs to an object or not.

(d) **Segmenting individual object instances:** It is a method used to distinguish an object one by one, as shown in Figure 1-3 (d). This method is used to distinguish between objects in an image.

Visual Scenes is used to create and manage online photo albums. It can be used to store photos on a personal website or blog or to share photos with friends and family. Visual Scenes provides an easy way to organize photos into albums, add captions and tags, and share photos with others.

Visual scenes is used to help people with autism spectrum disorder improve their communication and social skills. It is a software program that uses videos to help people understand facial expressions, body language, and social situations. The program is designed to help people with an autism spectrum disorder understand what is happening in a scene and learn how to respond to social situations.

Segmentation Problem

Synthetic data is important for machine learning, and especially for computer "vision." Synthetic data provides us with an almost unlimited amount of perfectly labeled data at a small cost. Computer scientists can create 3-D models of objects and their environments to create synthetic data with randomized conditions. This can help solve computer vision problems.

Semantic segmentation is the task of partitioning an image into semantically meaningful regions. Each pixel in the image is assigned a semantic label that reflects what the pixel represents. Semantic segmentation can be performed using a deep neural network with an encoder-decoder structure.

The segmentation problem in synthetic data is the problem of accurately dividing a dataset into homogeneous groups or segments. This is a difficult problem because it is often hard to determine what constitutes a homogeneous group and because there may be many ways to segment a dataset.

There are a few different approaches that can be used to segment a dataset. One approach is to use a clustering algorithm, which will group data points that are similar to each other. Another approach is to use a classification algorithm, which will assign each data point to a class. Finally, one can also use a rule-based approach, which will define a set of rules that determine how to segment the data.

The segmentation problem is particularly difficult in synthetic data because the data is often generated by a process that is not well understood. As a result, it is often hard to determine what features of the data are important for grouping the data points. In addition, synthetic data often contains a lot of noise, which can make it difficult to accurately segment the data.

Summary

In this chapter, you learned about synthetic data and why it is beneficial for data science and artificial intelligence. You also learned about the lifecycle of data and how synthetic data can be used to improve data quality at various stages. Finally, you learned how synthetic data is applied across various industries, including financial services, manufacturing, healthcare, automotive, robotics, security, social media, marketing, natural language processing, and computer vision.

Next, we'll begin delving into the different types of Synthetic Data.

References

[1]. G. Andrews, "What Is Synthetic Data?" Carrushome.com, 2022. https://www.carrushome.com/en/what-is-synthetic-data/ (accessed Apr. 13, 2022).

[2]. Scikit, "3.3. Metrics and scoring: quantifying the quality of predictions — scikit-learn 1.0.2 documentation." Sicikit Learn, 2022. https://scikit-learn.org/stable/modules/model_evaluation.html (accessed Apr. 13, 2022).

[3]. Airbus, "Can Synthetic Data Really Improve Algorithm Accuracy?" May 20, 2021. https://www.intelligence-airbusds.com/newsroom/news/can-synthetic-data-really-improve-algorithm-accuracy/, (accessed Apr. 13, 2022).spiepr Par10

[4]. S. Ramos et al., "RLDS: an Ecosystem to Generate, Share and Use Datasets in Reinforcement Learning," Nov. 2021, Accessed: Apr. 13, 2022. [Online]. Available: https://github.com/deepmind/envlogger.

[5]. N. Sambasivan, S. Kapania, H. Highfill, D. Akrong, P. Paritosh, and L. M. Aroyo, 'Everyone Wants To Do The Model Work, Not The Data Work': Data Cascades in High-Stakes AI, in Proceedings of the 2021 CHI Conference on Human Factors in Computing Systems, May 6, 2021, pp. 1–15, doi: 10.1145/3411764.3445518.spiepr Par109

[6]. G. Press, "Andrew Ng Launches A Campaign For Data-Centric AI," Forbes, Jun. 16, 2021. https://www.forbes.com/sites/gilpress/2021/06/16/andrew-ng-launches-a-campaign-for-data-centric-ai/?sh=d556d9d74f57, (accessed Apr. 13, 2022).

[7]. K. Singh, "Synthetic Data — key benefits, types, generation methods, and challenges," Towards Data Science, May 12, 2021. https://towardsdatascience.com/synthetic-data-key-benefits-types-generation-methods-and-challenges-11b0ad304b55 (accessed Apr. 13, 2022).

[8]. B. Ding et al., "DAGA: Data Augmentation with a Generation Approach for Low-resource Tagging Tasks," 2020. [Online]. Available: https://ntunlpsg.

[9]. MRP, "Natural Language Processing: Providing Structure To The Complexity Of Language." Jul. 30, 2018. https://www.mrpfd.com/resources/naturallanguageprocessing/ (accessed Jul. 02, 2022)

[10]. S. I. Nikolenko, Synthetic Data for Deep Learning, vol. 174. Cham:Springer International Publishing, 2021, doi: 10.1007/978-3-030-75178-4.

[11]. S. Colaner, "Why Unity claims synthetic data sets can improve computer vision models," VentureBeat, Jul. 17, 2022. https://venturebeat.com/2020/07/17/why-unity-claims-synthetic-data-sets-can-improve-computer-vision-models/ (accessed Apr. 13, 2022).

[12]. IBM, "What are Convolutional Neural Networks?," IBM Oct. 20, 2020. https://www.ibm.com/cloud/learn/convolutional-neural-networks (accessed Apr. 13, 2022)

[13]. J. Brownlee, "Overfitting and Underfitting With Machine Learning Algorithms," Machine Learning Algorithms, Aug. 12, 2019. https://machinelearningmastery.com/overfitting-and-underfitting-with-machine-learning-algorithms/ (accessed Apr. 13, 2022).

[14]. B. Marr, "Artificial Intelligence: The Clever Ways Video Games Are Used To Train AIs," Forbes, May 13, 2018. https://www.forbes.com/sites/bernardmarr/2018/06/13/artificial-intelligence-the-clever-ways-video-games-are-used-to-train-ais/?sh=5c45c30e9474 (accessed Apr. 17, 2022).

CHAPTER 2

Foundations of Synthetic data

This chapter explores the different types of synthetic data, how to generate fair synthetic data, and the benefits and challenges presented by synthetic data. It also explores the synthetic-real field gap and how to overcome it with field transfer, adaptation, and randomization. We discuss how simulation automates data labeling in autonomous vehicle companies and how real-world experience is inevitable. Finally, we discuss how data can be pretrained, learned with reinforcement, and self-supervised learning to learn medical images.

How to Generated Fair Synthetic Data?

"Fair synthetic data" refers to data that has been generated using causally-aware generative networks. These networks are designed to produce data that is free of bias and discrimination. This data can be used to train machine learning models that are fairer and more accurate. Additionally, this data can help reduce the overall amount of bias and discrimination in machine learning.

There are several reasons why it might be desirable to use fair synthetic data. For one, it can be difficult to obtain real-world data that is truly representative of the population. This is often due factors such as to self-selection bias, which can lead to certain groups being under-represented in the data. In addition, real-world data is often expensive to collect, so using synthetic data can be a more cost-effective option.

There are several ways to generate fair synthetic data. One approach is to use causally-aware generative networks. These networks are designed to learn the relationships between different variables in a data set and can than be used to generate new data that is consistent with these relationships. This ensures that the generated data is not biased concerning any sensitive attributes. This approach has several advantages

N. Gürsakal et al., *Synthetic Data for Deep Learning*, https://doi.org/10.1007/978-1-4842-8587-9_2

over other methods of generating synthetic data. First, it is much more efficient, as it only requires a single training step. Second, it is more accurate, as the generated data will be consistent with the known relationships between the variables. Finally, it is fairer, as the generated data will not biased concerning any sensitive attributes.

Generative models are a powerful tool for creating fair synthetic data, but they are not the only tool available. Other methods for creating synthetic data include data augmentation, which can be used to add diversity to a dataset, and resampling, which can be used to create balanced datasets. These methods can be used in conjunction with generative models to create even more realistic synthetic data.

One approach to generating fair synthetic data is to use data that has been anonymized or de-identified. This data has had all personally identifiable information removed, so it cannot be used to discriminate against individuals. However, this data may not be as realistic as data that has been generated using causally-aware generative networks. Another approach is to use data that has been generated using causally aware generative networks. This data is more realistic, but it may not be as anonymized or de-identified. Ultimately, the choice of method for generating fair synthetic data will depend on the needs of the user.

Now let's look at how to create synthetic data simply.

Generating Synthetic Data in A Simple Way

As shown in Figure 2-1, the synthetic data generation process is a three-step process that uses a statistical model to generate data that is statistically similar to real data. The first step is to create a model of the real data. The second step is to use the model to generate synthetic data. The third step is to compare the synthetic data to the real data to ensure that they are statistically similar. While generating synthetic data, a model is first fitted to the source data, and then synthetic data is generated by applying this model [2].

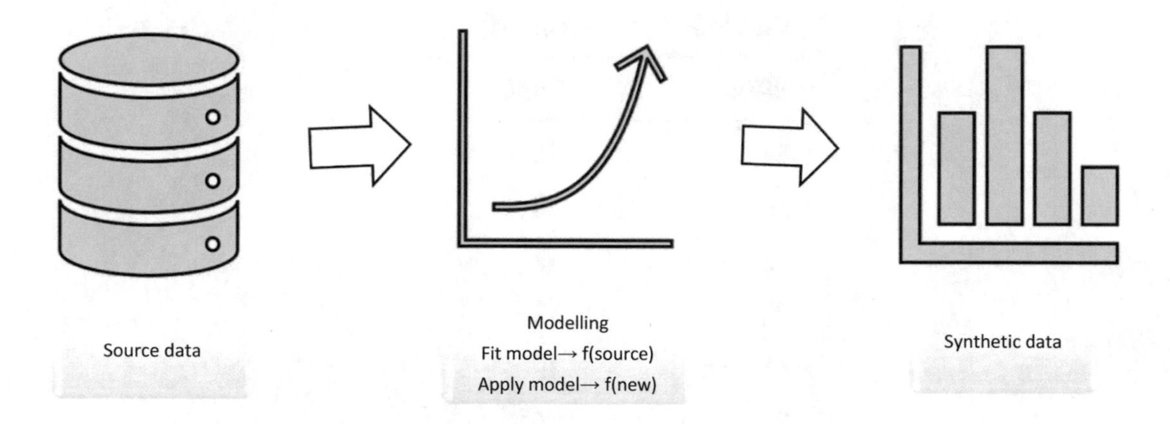

***Figure 2-1.** The synthetic data generation process*

Using the data in Table 2-1, we can generate synthetic data in a very simple way.

***Table 2-1.** Regression data*

DATA	TIME
3	1
5	2
4	3
7	4
5	5
8	6
12	7
10	8
11	9
14	10
16	11
12	12
13	13
20	14

(continued)

Table 2-1. *(continued)*

DATA	TIME
24	15
25	16
26	17
23	18
11	19
2	20
11	21
27	22
30	23
30	24
32	25

When the calculations are made, the following result is obtained:

```
Regression Equation

DATA = 2,35 + 1,025 TIME
```

FIT values are found for all observations by replacing the TIME variable in the regression equation. We can think that FIT values are synthetic data in a sense produced from DATA. When we think of the obtained FIT data as "data produced from data", these are referred to as "*augmented data*"; when we consider that these are only data produced from the found regression equation, they are close to the concept of "*synthetic data*". Another point to be noted in these results is that if we produce data with this regression equation, there will be a very large deviation from the real data in the 20th observations. The value of the 20th Observation for this DATA dataset is 2 and it is an outlier or a black swan. Although outliers represent very important turning points in real life, these data are difficult to produce as synthetic data. Figure 2-2 was obtained as a result of regression analysis.

```
Fits and Diagnostics for All Observations

Obs   DATA     Fit    Resid  Std Resid
  1   3,00    3,38    -0,38      -0,07
  2   5,00    4,40     0,60       0,11
  3   4,00    5,43    -1,43      -0,26
  4   7,00    6,45     0,55       0,10
  5   5,00    7,48    -2,48      -0,45
  6   8,00    8,50    -0,50      -0,09
  7  12,00    9,53     2,47       0,44
  8  10,00   10,55    -0,55      -0,10
  9  11,00   11,58    -0,58      -0,10
 10  14,00   12,60     1,40       0,25
 11  16,00   13,63     2,37       0,42
 12  12,00   14,65    -2,65      -0,47
 13  13,00   15,68    -2,68      -0,47
 14  20,00   16,71     3,29       0,58
 15  24,00   17,73     6,27       1,11
 16  25,00   18,76     6,24       1,10
 17  26,00   19,78     6,22       1,10
 18  23,00   20,81     2,19       0,39
 19  11,00   21,83   -10,83      -1,94
 20   2,00   22,86   -20,86      -3,75  R (Large residual)
 21  22,00   23,88    -1,88      -0,34
 22  27,00   24,91     2,09       0,38
 23  30,00   25,93     4,07       0,75
 24  30,00   26,96     3,04       0,56
 25  32,00   27,98     4,02       0,75
```

Regression Analysis: DATA Versus TIME

```
Analysis of Variance

Source      DF  Adj SS   Adj MS  F-Value  P-Value
Regression   1  1366,8  1366,84    40,69    0,000
  TIME       1  1366,8  1366,84    40,69    0,000
Error       23   772,6    33,59
Total       24  2139,4
```

```
Model Summary

      S    R-sq  R-sq(adj)  R-sq(pred)
5,79581  63,89%     62,32%      57,89%

Coefficients

Term        Coef  SE Coef  T-Value  P-Value   VIF
Constant    2,35     2,39     0,98    0,336
TIME       1,025    0,161     6,38    0,000  1,00

Regression Equation

DATA = 2,35 + 1,025 TIME

Fits and Diagnostics for Unusual Observations

Obs  DATA    Fit   Resid  Std Resid
 20  2,00  22,86  -20,86    -3,75  R

R  Large residual
```

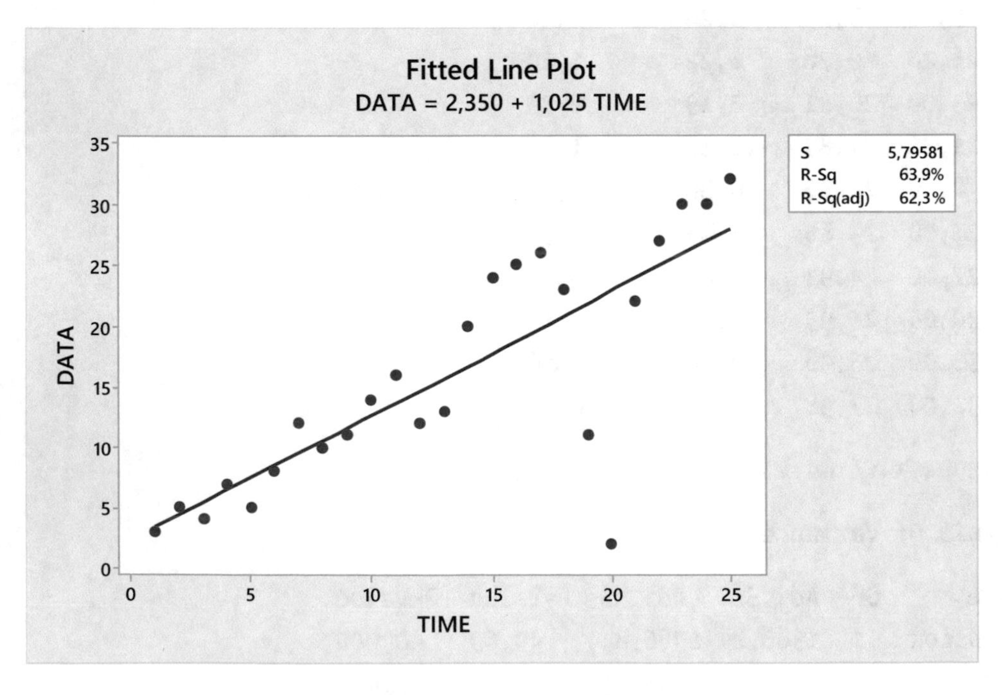

Figure 2-2. *The regression line for the data in Table 2-1*

The Black Swam problem in the context of AI systems is a major problem. AI systems are prone to unpredictable and unforeseeable events that can have a large or disproportional impact. Some examples of Black Swan events in the context of AI systems are:

- An AI system that suddenly starts acting out of character
- An AI system that becomes self-aware and decides to wipe out humanity
- A hacker taking over control of an AI system
- A natural disaster taking out an AI system
- A software bug causes an AI system to malfunction

The answer to this question will undoubtedly be unsupervised learning, whether supervised or unsupervised learning is more appropriate among the models of machine learning in the context of black swans.

Let's examine the use of synthetic data in video games.

Using Video Games to Create Synthetic Data

Researchers from Intel Labs and Darmstadt University in Germany have developed a way to automatically extract training data for machine learning from the computer game *Grand Theft Auto*. By using a software layer between the game and computer hardware, they were able to classify objects such as pedestrians, roads, and cars in the road images shown in the game [3].

The researchers found that commercial video games can be used "to create large-scale pixel-accurate ground truth data for training semantic segmentation systems" [4]. This allows for the development of more accurate systems for modern open-world games with realistic worlds, such as *Grand Theft Auto*, *Watch Dogs*, and *Hitman* [5]. Semantic segmentation datasets are collections of data that show how accurate different computer programs are at recognizing different objects and dividing them into different parts. They can help reduce the amount of hand-labeled real-world data needed because they show how accurate different computer programs are at recognizing different objects. Figure 2-3 shows that the approach can correctly label each pixel in the image, including the background, foreground, and object classes. The approach is also able to correctly label the boundaries between different objects and between different object classes.

Figure 2-3. *Semantic segmentation*

The relationship between AI and gaming has been longstanding and ever-growing. AI researchers have long seen games like chess as opportunities to test and showcase their work. Games provide an interesting and complex environment in which to test algorithms and AI. No matter how amazing your AI and machine learning algorithms are if you don't have enough high-quality data, all your work is useless. Donyaee Aram found that two factors were challenging his work on an AI-based computer vision solution that would allow the visually impaired to navigate around without obstructions and injury: the length of time required to collect the dataset and the size of the cost of tagging such a large dataset.

A dataset containing 213,400 images and automatically generated class annotations named SYNTHIA was generated by Ros et al. using Unity game engine [6]. SYNTHIA helps improve performance on the semantic segmentation task when used with real-world urban images that have been annotated manually [7] In a doctoral thesis in 2018 to generate synthetic street scene data, CoSy uses a 3D graphics approach to create convincing synthetic images to either replace or augment traditional data [6]. Also, in the CoSy system, randomized color selection can be made using the ColorModifier within the framework of domain randomization.

Driverless cars are now being trained by video game development machines so that autonomous cars can drive safely on the roads without crashing, like the studies in the examples we gave. These developments we are talking about brought to people's minds the thought that virtual datasets can be added to the datasets to be used for machine learning. In short, video game makers began providing computers with thousands of datasets in digital format so they could distinguish faces, products, landscapes, and anything that could be photographed. Later, Aram realized that when others faced the same problems, they used simulation instead of real data, creating tons of data in a very short time, at zero cost. There were many simulators for autonomous vehicles, such as the following [7]:

- CARLA Simulator
- AirSim (by Microsoft)
- Udacity Car Simulator
- Deepdrive
- Unity SimViz
- TORCS (The Open Racing Car Simulator)
- CitySim
- OpenAI Gym
- GSim
- DART

- SUMO
- ROS
- ENSA
- Drone Simulator
- EVE-NG
- MUSE

There were also simulators and datasets developed for the indoor movements of robots [8].

- Habitat AI (by Facebook)
- MINOS
- Gibson
- Replica
- MatterPort3D

Continuing her studies on the subject, Aram later determined that there is open-source software on GitHub and that the important data collection tools from the GTA V game are [8]:

- GameHook
- DeepGTAV
- DeepGTAV PreSIL [Lidar]
- GTAVisionExport

A computer scientist at the Xerox Research Center named Adrien Gaidon was surprised at how realistic it looked when he saw the trailer for the Assassin's Creed video game. He thought it was a movie trailer because of its lifelike graphics. But then he realized that it was all computer-generated imagery (CGI), and he was amazed that AI algorithms could be fooled in the same way. Next, Gaidon and his team began creating scenes used to train deep learning algorithms with Unity, a 3-D video game development machine. They not only created a synthetic environment but also started to transfer real images to the virtual environment. This allowed them to compare the effectiveness of algorithms trained on real images with those trained in virtual environments [9].

Grand Theft Auto can be used to learn "stop" signs in traffic. Artur Filipowicz and his team at Princeton University used an AI algorithm they wrote for the video game to learn the meaning of stop signs. They trained the AI algorithm with the appearance of partially covered, muddy, shaded, snow-covered stop signs at various times of the day. They used the video game *Grand Theft Auto V* as the training ground for the algorithm. Instead of a human playing the video game, they trained the AI algorithm by having it use a computer program [9].

Grand Theft Auto V allows you to use the footage for noncommercial purposes if you follow certain conditions. A way to get data out of the game is to use a middleware between the game and its graphics library. This can be done by detouring the library [10]. Detours is a library that allows you to inject code into arbitrary Win32 functions on x86 machines. It can intercept Win32 functions by replacing the target function images with their code [11].

Later, the technology in this field has developed in a way that detailed labeling can also be done automatically. We have found a way to automatically label things in videos using code from video games. This code comes from the Microsoft DirectX rendering API, which we use to inject specialized rendering code into the game as it is running. This code produces ground truth labels for things like segmenting objects, labeling their semantic meaning, estimating depth, tracking objects, and decomposing images into their intrinsic parts [12]. Researchers no longer need to laboriously label images; they can simply play video games all day long.

Synthetic data has now become a crucial component in the development of software to be used in autonomous vehicles, and Unity has gained expertise in this area. Unity allows you to test autonomous car software in a virtual environment, which provides several benefits. This is about how technology can be used to create things that wouldn't normally happen in the real world. For example, self-driving cars can be tested for millions of miles every day to make sure they are safe.

The importance of real-world data for training autonomous vehicles, robots, and other computer-based systems cannot be understated. However, synthetic data-data generated by computers-can also be useful. This is because it can be rich with variation due to what is known as "*limitless domain randomization*". This means that you can change colors, materials, and lighting within a simulation. You can use Unity to create a given simulation in which you can add multiple pedestrians and cars to an intersection and see how they interact with each other. This can be used to train a computer vision model [1]. For example, you can simulate car accidents or near misses.

Now let's explore Synthetic-to-Real Domain Gap.

The Synthetic-to-Real Domain Gap

The "synthetic-to-real domain gap" is a term used to describe the difference between the synthetic data use to train machine learning models and the real-world data that the models will be deployed on. This gap can lead to poor performance of machine learning models on real-world data, as the models have not been trained on data that is representative of the real world. Bridging the synthetic-to-real domain gap is therefore an important task to ensure that machine learning models can be deployed successfully on real-world data.

The synthetic-to-real domain gap can be caused by various factors, such as the different properties of the data in the two domains, the different distribution of the data in the two domains, or the different amounts of data in the two domains. This gap can be problematic for machine learning models because it can lead to overfitting of the synthetic data and poor performance on the real data. To address this issue, some methods have been proposed that aim to close the synthetic-to-real domain gap by either translating the synthetic data to the real domain or by training models on both synthetic and real data [13].

There are various ways of bridging the gap, which we'll explore below.

Bridging the Gap

Several methods aim to close the synthetic-to-real domain gap, but each has its advantages and disadvantages. Some methods translate the synthetic data to the real domain, while others train models on both synthetic and real data. Each method has its benefits and drawbacks, so it is important to choose the right method for the specific problem at hand.

- **Data augmentation** involves artificially generating new data that is similar to the existing data. This can be done by adding noise to the data or by using different algorithms to generate new data points. Data augmentation can help to close the synthetic-to-real domain gap by increasing the amount of data in the synthetic domain and by making the synthetic data more similar to the real data.
- **Transfer learning** involves training a model in one domain and then transferring the knowledge learned to another domain. This can be done by either fine-tuning a model that has been pre-trained on the

synthetic domain or by training a model on both the synthetic and real domains. Transfer learning can help to close the synthetic-to-real domain gap by making the synthetic data more like the real data.

- **Data preprocessing** involves transforming the data in the synthetic domain to make it more similar to the data in the real domain. This can be done by either normalizing the data or by using different algorithms to transform the data. Data preprocessing can help to close the synthetic-to-real domain gap by making the synthetic data more similar to the real data.
- **Ensembles** involve training multiple models on the same data and then combining the predictions of the models. This can be done by either training the models on different subsets of the data or by training the models on different types of data [14], [15]. Ensembles can help to close the synthetic-to-real domain gap by making the predictions of the models more robust and by making the synthetic data more similar to the real data.

Now let's briefly explain the subject of the domain transfer.

Domain Transfer

Synthetic data is commonly used in machine learning to train models that learn to generalize from data. A model trained on synthetic data can be used to make predictions on real data. Domain transfer is the process of using a model trained on synthetic data to make predictions on real data. This can be done by retraining the model on real data,or by using the model to generate synthetic data that is similar to the real data. Domain transfer is important because it allows models to be trained to data that is not available in the real world. This can be useful for training models that need to be robust on data that is not representative of the real world, such as data that is corrupted or missing.

Suppose you are going to put the autonomous vehicle you intend to produce for sale in San Francisco and Tokyo. The training data for the autonomous vehicle you will produce under these conditions must be obtained from San Francisco and Tokyo. If training data is only obtained from San Francisco and the vehicle is also sold in Tokyo, the vehicle's performance in Tokyo would be worse. In order to ensure that the autonomous vehicle performs well in both San Francisco and Tokyo, training data must be obtained from both cities. By obtaining training data from both San Francisco and Tokyo, the autonomous vehicle will be able to learn the different driving conditions in both cities and be able to perform well in both cities.

This time let's look at the subject in terms of robots. Deep reinforcement learning can help us learn a lot of data, but when transferring from a simulated environment to the real world, we often see a significant drop in performance. Therefore, it's important to have efficient transfer methods that close the gap between the two. Simulation-based training is less expensive than real-world training, but there are some differences between the two. One major difference is that simulation may not always match real-world settings. This can be a problem because it can be difficult to bridge the gap between simulation and reality. Methods that can account for mismatches in sensing and actuation are essential for making the most of simulation-based training. Reinforcement learning algorithms often use large amounts of labeled simulation data. However, the gap between simulation environments and the real world means that methods are needed to transfer knowledge learned in simulation to the real world [16]. The randomization algorithm is popular for sim-to-real transfer because it is effective in various tasks in robotics and autonomous driving. However, the reason why this algorithm works so well is not fully understood.

Domain transfer is the process of applying machine learning models trained on synthetic data to real-world data. This is often necessary because real-world data is often messy and difficult to work with, while synthetic data is cleaner and easier to use. By using domain transfer, we can improve the accuracy of our machine learning models and achieve better results in the real world. In this section, we will deal with two domain transfer problems such as domain adaptation and domain randomization.

Now Let's briefly explain the subject of domain adaptation.

Domain Adaptation

In a machine learning setting if we want to make a disease diagnosis and if x is the properties of the patient, then $y=f(x)$ is a disease. In a face recognition problem, if x is a bitmap picture of a person's face; then $y=f(x)$ is the name of that person. For an autonomous car, if x is a bitmap picture of the road surface in front of a car, then $y=f(x)$ is the degree to turn the steering wheel. In all these problems, we know x and y pairs, but we don't know the function f, and we try to find this function. At the forefront of machine learning problems, we slice our data set as training and test set. Besides we want our test set to be large enough to produce statistically meaningful results, and it should be representative of the data set so that the results are not skewed [17]. Computer vision applications, such as face recognition, must be able to adapt to the specific distribution of data in each domain. For example, the distribution of faces in a crowd will be different from the distribution of faces in a family photo.

The assumption in classical machine learning is that the distributions of the training and test sets are the same. If this assumption is correct, then the model learned with the labeled training data should work well on the test data as well. However, in real life, the training and test sets may be collected from different sources for various reasons, and this assumption may not be correct. In this case, the learned model may not work well on the test data.

If certain events occur more frequently in the data, then the data is biased. For example, data collected from older people relative to the total population is biased. The model we found for biased data would consider certain outcomes to occur with greater probability. And if we had to give an example, it might consider certain levels of blood pressure to be normal, when they would indicate a health risk for younger patients.

If we want our classification function to make accurate predictions for new samples, we must use a dataset that is representative of the distribution from which the samples are drawn. However, if our data is not randomly sampled from the population, our training and test data will be different. In this case, standard classifiers will not perform well. This is because the classifier will not be able to learn the true underlying distribution of the data.

There are two main ways to solve the problem of differences between the distributions of training and test data: domain adaptation and transfer learning. Domain adaptation focuses on how to make the training data more similar to the test data, while transfer learning focuses on how to use knowledge from one domain to another. Both of these concepts focus on how to solve the problem of disparity between distributions of training and test data.

Domain adaptation is a technique that is used to improve the performance of a machine learning algorithm when the data is collected from a different source domain than the one for which the algorithm was originally trained. This is often done by adjusting the algorithm to better fit the new data distribution. Transfer learning is a technique that is used to improve the performance of a machine learning algorithm [18] when the data is collected from a different target domain than the one for which the algorithm was originally trained. This is often done by adjusting the algorithm to better fit the new target data distribution. Domain adaptation is the ability to apply an algorithm that is trained on one or more source domains to a different target domain [19] so that the features extracted from the source and target dataset are similar. This is useful for tasks such as image recognition, where the features in the source and target datasets may be different.

Domain adaptation is a problem of accurately transferring knowledge from a training dataset to a test dataset when the distributions of the two datasets are different. Some proposed solutions to this problem are to use a mix of real and synthetic data, or to use a domain-specific language model.

One of the most important challenges in machine learning is to develop methods that can learn from data in a new domain, i.e., to adapt the learned model to the new domain. This is known as *domain adaptation*. Domain adaptation is important because the data in different domains can be quite different in terms of structure, distribution, and noise. It is also important in many practical applications such as cross-domain learning, where the task is to learn a model that can be applied to a new domain, or transfer learning, where a model is pre-trained on a large dataset and then used to learn a new task in a different domain.

Domain adaptation has been studied in a variety of different ways, including:

- Data augmentation, where additional data is artificially generated to make the data in the new domain more similar to the data in the source domain.
- Algorithmic adaptation, where the adaptation algorithm is specifically designed for the new domain.
- Feature adaptation, where the features used by the adaptation algorithm are adapted to the new domain; and
- Algorithmic enhancement, where the adaptation algorithm is enhanced in some way for the new domain.

One of the most common ways of adapting a machine learning algorithm is to use data augmentation. This involves artificially generating additional data to make the data in the new domain more similar to the data in the source domain. This can be done in several ways, including:

- Binarization, where the data is converted to binary form.
- Noise addition, where random noise is added to the data.
- Transformations, where the data is transformed in some way, such as by scaling, translation, or rotation; and

- Generative models, where a generative model is used to generate additional data.

Algorithmic adaptation is another way of adapting a machine learning algorithm for a new domain. In this approach, the adaptation algorithm is specifically designed for the new domain. This can be done in several ways, including:

- Algorithmic modification, where the adaptation algorithm is modified to make it more suitable for the new domain.
- Algorithmic combination, where two or more adaptation algorithms are combined to form a new adaptation algorithm; and
- Algorithmic optimization, where the adaptation algorithm is optimized to make it more suitable for the new domain.

Feature adaptation is another way of adapting a machine learning algorithm for a new domain. In this approach, the features used by the adaptation algorithm are adapted to the new domain. This can be done in several ways, including:

- Feature selection, where the features used by the adaptation algorithm are selected based on their suitability for the new domain.
- Feature extraction, where the features used by the adaptation algorithm are extracted from the data in the new domain.
- Feature transformation, where the features used by the adaptation algorithm are transformed to make them more suitable for the new domain; and
- Feature combination, where two or more features are combined to form a new feature.

CNNs (convolutional neural networks) are being successfully used in industrial businesses such as autonomous driving and medical imaging for semantic segmentation. However, the large amount of data required for CNNs has been a problem for collecting and labeling data. However, recent technological developments have made it possible for CNN models to produce and label photo-like synthetic data. CNNs are effective at semantic segmentation of urban scenes, reducing the domain mismatch between real and synthetic images. A curriculum-style learning approach has been proposed to further improve results [20].

Now let's examine domain randomization in terms of synthetic data.

Domain Randomization

Domain randomization is a technique for training neural networks that can improve classification performance by making the model more robust to domain shifts. In this figure, the input data is randomly perturbed, which makes the model more resistant to changes in the data distribution. This can be helpful when there is a mismatch between the training data and the test data.

One of the biggest problems when training artificial neural networks is the large data requirements of these networks. In applications such as CNN, although tens of thousands of images may be pre-labeled manually or automatically, the results obtained for data reserved for testing purposes may not be generalizable. With synthetic image datasets produced using domain randomization, it is possible to obtain accuracy rates close to real data, such as 88%. Domain randomization is a technique that is used to help an AI model understand the general pattern by using thousands of variations of an object and its environment. This allows the AI model to better understand real data when it is applied. In a summary, domain randomization is about learning a model by seeing the objects in different lights, and colors, in a different order, in short, in different environments.

Synthetic data can differ significantly from real data in several ways, such as the distribution of light and patterns. This often creates a "reality gap" between the two types of data. If domain randomization is applied and synthetic data is generated with a sufficiently large amount of variability, the reality gap is reduced. While applying domain randomization, synthetic data is tried to be produced in a wide variety of conditions, considering the angle, light, pattern, structure, shadow, and camera effects. In artificial neural networks trained on a sufficiently wide variety of synthetically generated data, the learned models can be generalized to real data (real objects). Thus, the essence of the concept of domain randomization is the generation of "a sufficiently wide variety" of synthetic data. Domain randomization is a technique used to reduce the reality gap. By randomly selecting different simulated environments to train the model in, the model is exposed to a greater variety of scenarios. This helps to ensure that the model performs well in a wider range of situations.

Domain randomization is a process that is used to create synthetic datasets. These datasets are used to train machine learning models for tasks such as cat vs. dog classification. The accuracy levels of synthetic datasets can reach up to 88% [21]. This is an important tool for machine learning researchers, as it allows them to train models using data that is more representative of the real world. In addition, randomization can be used to create more diverse datasets, which can help to improve the generalizability of models.

A deep neural network system for object detection can be improved by using synthetic images that are generated with variability. This is done by using the domain randomization technique. Domain randomization is a technique used in machine learning to force the neural network to learn the essential features of the object of interest. In this technique, the simulator's lighting, pose textures, etc. parameters are randomized in non-realistic ways [22]:

In the domain randomization technique, the domain of an image can be randomized in several ways. Some common methods are to randomize the color palette, the image content, or the image size. Another method is to randomly change the image compression algorithm, which can result in a significantly different image file size. Another method is to randomly select a different image processing filter to use on the image. This can result in a completely different look for the image. Finally, the image can be randomized by changing the order of the pixels in the image file.

Is Real-World Experience Unavoidable?

According to Jonny Dyer, director of engineering in Lyft's Level 5 self-driving division [23], simulations are not perfect, and they will never completely replace real-world experience. However, they are a valuable tool in helping us to develop self-driving technology. By calibrating and validating simulations against reality, we are making great progress in our ability to create self-driving cars. Simulations can help us to learn about and test different aspects of self-driving technology, including how the algorithms that control the cars respond to different situations. They can also help us to identify potential issues and problems before they occur in the real world. By using simulations, we can make self-driving cars that are much safer and more reliable.

Nowadays, both artificial and real data are being used together in machine learning applications. Most of the data used are synthetic, but some are real. This helps to improve the accuracy of the machine learning applications. Waymo CEO John Krafcik has said that real-world experience is necessary for developing driverless cars [23]. Also, Yandex has a similar view. According to Yandex, which operates autonomous cars in Moscow, the simulation benefits the development of these cars, but testing in the public domain is critical. Without real-world testing, autonomous car companies can't get real driving data. Doing this work without fully testing simulation programs on real roads will prevent the development of autonomous vehicles in the short term. Some experts believe that a combination of both real and simulated data is necessary for developing driverless cars. Real-world experience is necessary to test the capabilities of the cars, while simulated data can be used to improve the accuracy of the machine learning applications. Without both, it would be difficult to develop autonomous vehicles in the short term.

Now let's briefly explain the subject of pretraining.

Pretraining

Because it requires a lot of data and labeling this data, we know that supervised learning is a bottleneck. In self-supervised learning, which can be used to overcome this bottleneck, important features of the data are determined with the help of a model using some of the unlabeled data used in learning. Although the starting point here is unlabeled data, real data is still required for such applications. "Self-supervised learning is to predict any unobserved or hidden part (or property) of the input from any observed or unhidden part of the input" and it is known to accelerate apps like proactive detection of hate speech on Facebook [24].

Pretraining is a process whereby a deep learning model is first trained on a large dataset, typically using a supervised learning technique, before being fine-tuned on a smaller dataset. The large dataset is typically a labeled dataset that is similarly to the smaller dataset that the model will be fine-tuned on. The aim of pretraining is to provide the deep learning model with initial weights that are close to the optimum weights for the task at hand. This can save a lot of time and computational resources when training the model on the smaller dataset as the model will not need to learn from scratch.

Pretraining is a popular approach in deep learning as it can yield very good results with relatively little data. It is especially useful when labeled data is scarce or expensive to obtain. In these cases, pretraining can be used to train a model on a large dataset that is readily available, such as an unlabeled dataset. The model can then be fine-tuned on the smaller labeled dataset.

Pretraining can be done using a variety of methods, including self-supervised learning, unsupervised learning, and transfer learning [25]. Self-supervised learning is a technique whereby the model is trained to predict labels from data that is not labeled. This can be done, for example, by training the model to predict the next frame in a video or the next word in a sentence. Unsupervised learning is a technique whereby the model is trained on data that is not labeled and is not required to make predictions. Transfer learning is a technique whereby a model that has been trained on one task is used to initialize the weights of a model that is being trained on a different task.

Pretraining is an important technique in deep learning as it can provide the model with good initial weights that are close to the optimum weights for the task at hand. This can save a lot of time and computational resources when training the model on a smaller dataset. In pretraining applications with synthetic data against self-supervised learning, the need for real data is no longer required. For example, in information retrieval (IR) applications made with neural networks, it was seen that pre-training with synthetic data improves retrieval performance. Some authors have also used synthetic images generated via graphics engines for effective pretraining and explained their results as, "While a large portion of contemporary representation learning research focuses on self-supervision to avoid using labels, we hope our demonstration with Task2Sim motivates further research in using simulated data from graphics engines for this purpose" [26]. By controlling the simulation parameters (lighting, pose, materials, etc.) with the graphics engines used in such applications, a pretraining dataset is created and can be used for learning purposes. The Unity Perception package provides a randomization framework that simplifies introducing variation into synthetic environments and includes a set of labelers. For example, a synthetic aerial image dataset for ship recognition, called UnityShip has been proposed. "This dataset contains over 100,000 synthetic images and 194,054 ship instances, including 79 different ship models in ten categories and six different large virtual scenes with different periods, weather environments, and altitudes" [27] and it has been found "that for small- and medium-sized real-world datasets, the synthetic dataset provides a large improvement in model pre-training and data augmentation" [27]. One of the results obtained in synthetic data generation research is to achieve a high degree of realism with computer graphics.

Now let's examine Reinforcement Learning in terms of synthetic data.

Reinforcement Learning

In machine learning we know that supervised learning requires labeled data, and unsupervised learning requires datasets that do not require labels. When we consider reinforcement learning, this technique is close to unsupervised learning as the data used while training the machine is not labeled. However, because reinforcement learning is trained through the reward and punishment mechanism of an agent, it is not well suited to the classification of supervised and unsupervised learning. GAN models can be used to generate synthetic data that can be used to train medical models [28]. This data can improve the accuracy of medical models and help be used to reduce the training time required. Additionally, synthetic data can be used to select images that are more likely to contain informative features.

In reinforcement learning, the agent tries to find the best way to achieve a goal by learning through trial and error. Actions that lead to better results receive higher rewards, while actions that lead to worse results receive lower rewards or even punishments (negative rewards). The agent's policy is the strategy it uses to select actions in each state. By interacting with the environment, the reinforcement learning system learns to solve a given task by finding an action selection policy that maximizes the accumulated reward. The agent continues to learn and update its policy as it interacts with the environment, to find the best possible strategy for achieving the goal. Over time, the reinforcement learning system should be able to solve the given task more efficiently and effectively.

Reinforcement learning is used in NLP to improve text summarization, question answering, and translation. This is done by giving feedback to the system on how well it is doing at a task. Reinforcement learning can be used to improve the performance of autonomous vehicles in areas such as trajectory optimization, motion planning, dynamic path determination, and controller optimization. DeepRacer is a self-driving car that has been designed to test out reinforcement learning on a physical track.

An autonomous vehicle is using reinforcement learning when it is learning how to complete a task by being rewarded for completing it correctly. This type of learning is done by giving the vehicle positive reinforcement (such as a treat) when it completes a task correctly and negative reinforcement (such as noise) when it completes a task incorrectly. This allows the vehicle to learn which tasks are correct and which are incorrect and allows it to improve its performance over time.

One advantage of using reinforcement learning is that it can be used to optimize things without human intervention. This can be useful for tasks that are difficult or impossible for humans to do, such as optimizing a robot's grasping behavior. Additionally, reinforcement learning can be used to learn from experience, which can make it more efficient than other methods.

There is a technique called reinforcement learning that can be used to optimize things in a simulated environment. For example, Facebook has developed a program called Horizon that uses reinforcement learning to help robots do tasks better, like grasping. This approach is effective in many different domains, from video games to industrial robotics.

Finally, let's examine self-supervised learning.

Self-Supervised Learning

One common challenge in many medical applications is the scarcity of labeled data. However, unlabeled data is readily available in many medical domains. Self-supervised pretraining can be very helpful in applications where labeled data is scarce. CNNs can transfer well between tasks, meaning that they can be pre-trained on a task and then applied to a new task with a modest amount of labeled data [29]. In this approach, the first step is to train a deep learning model on a dataset of labeled images, such as ImageNet. This tunes the parameters of the model's layers to the general patterns found in all kinds of images.

Next, we have a trained deep learning model that can be fine-tuned on a limited set of labeled examples for the target task. However, it should be accepted that medical images are very different from natural images in ImageNet, and so the deep learning model may not be as accurate when used to classify medical images [30].

The research team at Google used a variation of the SimCLR framework called Multi-Instance Contrastive Learning to create representations of images [31] that are stronger than if they had only been used in supervised learning. MICLe is a framework for learning from multiple images of the same patient, even if the images are not annotated for supervised learning [30]. This means that the Google team used a computer program to learn from unlabeled images. The program was able to learn how to recognize patterns in the images and create representations of them that were stronger than if the team had only used supervised learning.

In the SimCLR framework called MICLe, the Google Research team found that the composition of data augmentations (adding new data to help improve predictions) plays a critical role in defining effective predictive tasks. They also found that introducing a learnable nonlinear transformation (a change in how data is represented) between the representation and the contrastive loss (the measure of how well predictions match the actual data) substantially improves the quality of the learned representations. Additionally, contrastive learning (learning how to predict differences between data sets) benefits from larger batch sizes and more training steps compare to supervised learning (learning how to predict specific values within data sets). Finally, they found that they could outperform previous methods for self-supervised and semi-supervised learning [32] (learning without complete data sets) on the ImageNet data set by combining these findings.

In the medical imaging field one of the biggest challenges in medical imaging is the lack of data. So, researchers have turned to use GANs to generate fake medical images. This makes the data set bigger, and in turn, helps improve the accuracy of CNNs when it comes to classifying images. This is especially useful in cases where real data is difficult or impossible to obtain, such as in rare disease diagnoses. GANs have also been used to generate realistic-looking images of organs and tissues, which can be used to train CNNs for medical image segmentation. Overall, GANs can be a powerful tool in the medical imaging field, helping to improve the accuracy of CNNs and providing realistic images for training and testing.

Self-supervised learning is a type of machine learning where the model is trained using data that is not labeled. This can be done in several ways, such as using data augmentation to create new data points from the original data or using unsupervised learning algorithms to learn from the data [33].

Self-supervised learning has several advantages over traditional supervised learning. First, it is more efficient because it does not require labels. Second, it can be used to learn from data that is not easily labeled, such as images or videos. Finally, self-supervised learning can be used to learn features that are not easily learned with supervised learning, such as high-level features in images or videos.

Self-supervised learning is a powerful tool for training deep neural networks. However, it is important to remember that self-supervised learning is not a replacement for supervised learning. Supervised learning is still necessary for tasks that require labels, such as classification.

Summary

In this chapter, you learned about the different types of synthetic data and how they can be used to generate fair synthetic data. You also learned about how to use video games to create synthetic data, as well as the synthetic-to-real domain gap. Additionally, you learned about domain transfer, domain adaptation, and domain randomization. Finally, you learned about whether the real-world experience is unavoidable, and about pretraining, reinforcement learning, and self-supervised learning.

Next, we'll begin delving into Introduction to GANs.

References

[1]. S. Colaner, "Why Unity claims synthetic data sets can improve computer vision models," VentureBeat, Jul. 17, 2022. https://venturebeat.com/2020/07/17/why-unity-claims-synthetic-data-sets-can-improve-computer-vision-models/ (accessed Apr. 13, 2022).

[2]. Y. Hilpisch, Artificial Intelligence in Finance. O'Reilly Media. 2020.

[3]. W. Knight, "Self-Driving Cars Can Learn a Lot by Playing Grand Theft Auto," MIT Technology Review, Sep. 12, 2016. https://www.technologyreview.com/2016/09/12/157605/self-driving-cars-can-learn-a-lot-by-playing-grand-theft-auto/ (accessed Apr. 17, 2022).

[4]. A. Kumar, "Deep-Learning Features, Graphs And Scene Understanding," International Institute of Information Technology, Hyderabad, 2020. Accessed: Apr. 17, 2022. [Online]. Available: http://cdn.iiit.ac.in/cdn/cvit.iiit.ac.in/images/Thesis/MS/Abhijeet_kumar/Abhijeet_kumar_thesis.pdf

[5]. S. R. Richter, V. Vineet, S. Roth, V. Koltun, and T. U. Darmstadt, "Playing for Data: Ground Truth from Computer Games," 2016.

[6]. N. Bhandari, "Procedural synthetic data for self-driving cars using 3D graphics," Massachusetts Institute of Technology, 2018.

[7]. G. Ros, L. Sellart, J. Materzynska, D. Vazquez, and A. M. Lopez, "The SYNTHIA Dataset: A Large Collection of Synthetic Images for Semantic Segmentation of Urban Scenes," in Proceedings of the IEEE Conference on Computer Vision and Pattern Recognition (CVPR), 2016, pp. 3234–3243.

Accessed: Apr. 17, 2022. [Online]. Available: https://www.cv-foundation.org/openaccess/content_cvpr_2016/papers/Ros_The_SYNTHIA_Dataset_CVPR_2016_paper.pdf

[8]. A. Donyaee, "How video games can provide quality data for AI applications," Towards Data Science, May 20, 2020. https://towardsdatascience.com/i-play-video-games-but-not-for-entertainment-c20d28d998bf (accessed Apr. 17, 2022).

[9]. B. Marr, "Artificial Intelligence: The Clever Ways Video Games Are Used To Train AIs," Forbes, May 13, 2018. https://www.forbes.com/sites/bernardmarr/2018/06/13/artificialintelligence-the-clever-ways-video-games-are-used-to-train-ais/?sh=5c45c30e9474 (accessed Apr. 17, 2022).

[10]. S. R. Richter, T. U. Darmstadt, Z. Hayder, and V. Koltun, "Playing for Benchmarks," in Proceedings of the IEEE International Conference on Computer Vision., 2017, pp. 2213–2222.

[11]. G. Galen Hunt and D. Brubacher, "Detours: Binary Interception of Win32 Functions," in Proceedings of the 3rd USENIX Windows NT Symposium, 1999, pp. 135–144. Accessed: Apr. 17, 2022. [Online]. Available: https://www.usenix.org/legacy/events/usenixnt99/full_papers/hunt/hunt_html/.

[12]. P.Krähenbühl, "Free supervision from video games," in Proceedings of the IEEE Conference on Computer Vision and Pattern Recognition (CVPR), 2018, pp. 2955–2964.

[13]. R. Liu, C. Yang, W. Sun, X. Wang, and H. Li, "StereoGAN: Bridging Synthetic-to-Real Domain Gap by Joint Optimization of Domain Translation and Stereo Matching," May 2020., [Online]. Available: http://arxiv.org/abs/2005.01927.

[14]. R. Jin, J. Zhang, J. Yang, and D. Tao, "Multi-Branch Adversarial Regression for Domain Adaptative Hand Pose Estimation," IEEE Transactions on Circuits and Systems for Video Technology, pp. 1–1, 2022, doi: 10.1109/TCSVT.2022.3158676.

[15]. S. Zhao et al., "A Review of Single-Source Deep Unsupervised Visual Domain Adaptation," IEEE Transactions on Neural Networks and Learning Systems, vol. 33, no. 2, pp. 473–493, Feb. 2022, doi: 10.1109/TNNLS.2020.3028503.

[16]. W. Zhao, J. P. Queralta, and T. Westerlund, "Sim-to-Real Transfer in Deep Reinforcement Learning for Robotics: a Survey," in 2020 IEEE Symposium Series on Computational Intelligence (SSCI), Dec. 2020, pp. 737–744, doi: 10.1109/SSCI47803.2020.9308468.

[17]. Google Developers, "Training and Test Sets: Splitting Data," Machine Learning Crash Course, 2020. `https://developers.google.com/machine-learning/crash-course/training-and-test-sets/splitting-data` (accessed Apr. 17, 2022).

[18]. A. Descals, S. Wich, E. Meijaard, D. L. A. Gaveau, S. Peedell, and Z. Szantoi, "High-resolution global map of smallholder and industrial closed-canopy oil palm plantations," Earth System Science Data, vol. 13, no. 3, pp. 1211–1231, Mar. 2021, doi: 10.5194/essd-13-1211-2021.spiepr Par134

[19]. V. Lendave, "Understanding Direct Domain Adaptation in Deep Learning," Aug. 28, 2021. `https://analyticsindiamag.com/understanding-direct-domain-adaptation-in-deep-learning/` (accessed Apr. 17, 2022).

[20]. Y. Zhang, P. David, and B. Gong, "Curriculum Domain Adaptation for Semantic Segmentation of Urban Scenes," in 2017 IEEE International Conference on Computer Vision (ICCV), Oct. 2017, pp. 2039–2049, doi: 10.1109/ICCV.2017.223.

[21]. S. Z. Valtchev and J. Wu, "Domain randomization for neural network classification," Journal of Big Data, vol. 8, no. 1, p. 94, Dec. 2021, doi: 10.1186/s40537-021-00455-5.

[22]. J. Tremblay, T. To, and S. Birchfield, "Falling Things: A Synthetic Dataset for 3D Object Detection and Pose Estimation Thang To NVIDIA," in Proceedings of the IEEE Conference on Computer Vision and Pattern Recognition Workshops, 2019, pp. 2151–2154. Accessed: Apr. 15, 2022. [Online]. Available: `http://research.`

[23]. K. Wiggers, "The challenges of developing autonomous vehicles during a pandemc," VentureBeat, Apr. 28, 2020. https://venturebeat.com/2020/04/28/challenges-of-developing-autonomous-vehicles-during-coronavirus-covid-19-pandemic/ (accessed Apr. 18, 2022).

[24]. Y. LeCun and I. Misra, "Self-supervised learning: The dark matter of intelligence," Mar. 04, 2021. https://ai.facebook.com/blog/self-supervised-learning-the-dark-matter-of-intelligence/ (accessed May 24, 2022).

[25]. J. Sun, D. Wei, K. Ma, L. Wang, and Y. Zheng, "Boost Supervised Pretraining for Visual Transfer Learning: Implications of Self-Supervised Contrastive Representation Learning," 2022. Accessed: May 24, 2022. [Online]. Available: www.aaai.org

[26]. S. Mishra et al., "Task2Sim: Towards Effective Pre-training and Transfer from Synthetic Data," 2022. [Online]. Available: https://samarth4149.github.io/projects/task2sim.html

[27]. B. He, X. Li, B. Huang, E. Gu, W. Guo, and L. Wu, "UnityShip: A Large-Scale Synthetic Dataset for Ship Recognition in Aerial Images," Remote Sensing, vol. 13, no. 24, p. 4999, Dec. 2021, doi: 10.3390/rs13244999.

[28]. J. Drechsler, Synthetic datasets for statistical disclosure control: theory and implementation. Springer, 2011. Accessed: Apr. 15, 2022. [Online]. Available: https://books.google.com/books/about/Synthetic_Datasets_for_Statistical_Discl.html?hl=tr&id=RYisNAEACAAJ.

[29]. Y. M. Asano, C. Rupprecht, and A. Vedaldi, "A critical analysis of self-supervision, or what we can learn from a single image," 2020. https://arxiv.org/abs/1904.13132.

[30]. B. Dickson, "Google Research: Self-supervised learning is a game-changer for medical imaging," TechTalks, Nov. 08, 2021. https://bdtechtalks.com/2021/11/08/google-research-self-supervised-learning-medical-imaging/ (accessed Apr. 18, 2022).

[31]. C. Chen, X. Zhao, and M. C. Stamm, "Generative Adversarial Attacks Against Deep-Learning-Based Camera Model Identification," IEEE Transactions on Information Forensics and Security, pp. 1–1, 2019, doi: 10.1109/TIFS.2019.2945198.

[32]. T. Chen, S. Kornblith, M. Norouzi, and G. Hinton, "A Simple Framework for Contrastive Learning of Visual Representations," in Proceedings of the 37th International Conference on Machine Learning, PMLR, 2020, pp. 1597–1607. Accessed: Apr. 18, 2022. [Online]. Available: `https://github.com/google-research/simclr`.

[33]. J. Arun Pandian, G. Geetharamani, and B. Annette, "Data Augmentation on Plant Leaf Disease Image Dataset Using Image Manipulation and Deep Learning Techniques," in 2019 IEEE 9th International Conference on Advanced Computing (IACC), Dec. 2019, pp. 199–204, doi: 10.1109/IACC48062.2019.8971580.

CHAPTER 3

Introduction to GANs

This chapter provides an introduction to the basic concepts of *generative adversarial networks* (GANs), which are a type of neural network that can be used to generate synthetic data. Later chapters demonstrate how to use R and Python to use GANs to generate synthetic data for specific applications; this chapter provides crucial context for the R and Python chapters on generating synthetic data.

After brief introduction to GANs, we'll explore the different types, such as CTGAN, SurfelGAN, CycleGAN, SinGAN, MedGAN, DCGAN, WGAN and SeqGAN of GANs available. Next we will cover, Conditional GANs and BigGAN.

GANs

GANs are a type of neural network that was created in 2014 by Goodfellow et al. [1]. They are composed of two separate neural networks: the generator and the discriminator. These networks are designed to compete against each other to fool each other. To build a GAN model, you need to create the generator and discriminator networks and combine them into a GAN model. GANs are used to generate high-dimensional data. The first network, called the generator, learns to generate data that is likely to be distributed in a certain way. The second network, called the discriminator, tries to classify examples as either real or fake. The discriminator is trained to be good at distinguishing real data from fake data.

GANs have found success in recent years in tasks such as image generation, where they can create realistic images that are even to the point of being indiscernible from reality. GANs are computer programs that can be used to create realistic images. They successful at translating photos of summer to winter or day to night and generating realistic photos of objects, scenes, and people [2]. And we should point out that, GANs are less prone to over-fitting. GANs have been used to generate images of people who

N. Gürsakal et al., *Synthetic Data for Deep Learning*, https://doi.org/10.1007/978-1-4842-8587-9_3

don't exist and to creat *deepfakes*-videos in which people's face are superimposed onto other people's bodies.

GANs and variational autoencoders are two types of deep generative models. A variational autoencoder is a deep generative model that has an encoder, a decoder, and a loss function, as shown in Figure 3-1. The encoder takes an input and turns it into a representation. The decoder takes the representation and turns it back into the original input. The loss function measures how close the output is to the original input.

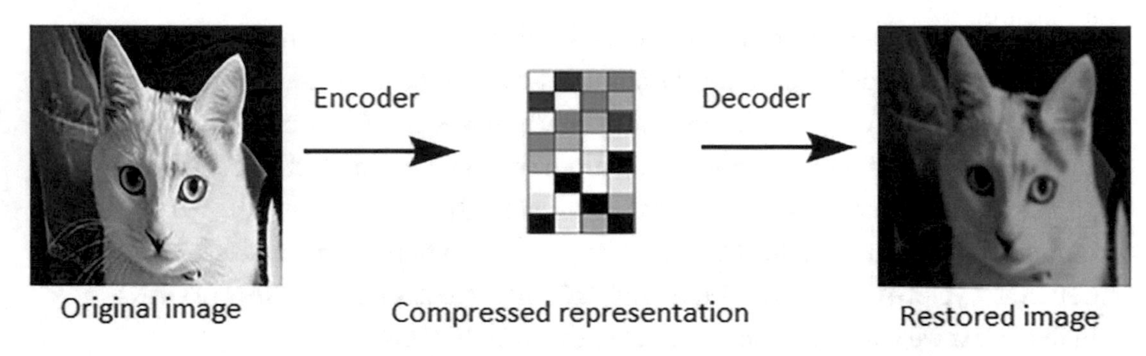

Figure 3-1. *A simple autoencoder structure*

The most important feature of autoencoders is that they can learn how to compress data without any supervision. This is a valuable ability because it means that autoencoders can be used to learn how to represent data in a more efficient way, which can lead to better performance on tasks such as classification and prediction.

Autoencoders are a type of neural network used for learning compressed representations of data. They are typically composed of an encoder network and a decoder network, which are connected in a feedback loop. The encoder network takes an input vector and produces a compressed representation, while the decoder network takes the compressed representation and produces the original input vector.

An autoencoder is like Principal Component Analysis but can learn a non-linear manifold. An autoencoder can be trained either end-to-end or layer by layer. When it is trained layer by layer, it is "stacked" together, which leads to a deeper encoder.

VAE-GANs are a hybrid of GANs and variational autoencoders. They are typically superior to deep generative models and are more difficult to work with than variational autoencoders. However, they require a lot of data and tuning to achieve good results encoders can be used to learn an "informative" representation of data that can be used for various purposes, such as clustering. VAE-GANs are used in a variety of "fields, including computer vision, natural language processing, and machine learning". For example, text recognition, image recognition, face recognition, and speech recognition also use VAE-GANs.

GANs can generate data that is more realistic than VAEs, but they require more time to do so. VAEs smooth the edges of images, while GANs generate sharper images. The main difference that distinguishes GANs from other synthetic data generation models is that it is a type of synthetic data generation model that can learn the real data distribution of a training data set and generate new data points with some variation from this distribution.

GANs are a type of algorithm used to generate artificial data. There are many variants of GANs, each with its strengths and weaknesses. The Wasserstein GAN, for example, is a variant that uses the Wasserstein distance to train the GAN. This has many benefits, such as improved stability of learning and the ability to avoid problems like mode collapse.

Now let's give information about the types of GANs. Let's take CTGAN first.

CTGAN

The CTGAN (Convolutional Tensor-based Generative Adversarial Network) algorithm was first proposed by Chen et al. in 2017 [3]. The original CTGAN algorithm was improved by Zhang et al. in 2018 [4]. The Zhang et al. algorithm is called the Dual-GAN (DGAN). The DGAN algorithm can learn a wider range of image transformations and produces more realistic images.

GANs have been successfully used to generate realistic images [5]. The CTGAN is a recent development in the field of GANs, which uses a CNN to learn the low-level features of the data, and a GAN to learn the high-level features. The advantage of the CTGAN is that it can generate realistic images with much more detail than a standard GAN.

The CTGAN algorithm is a GAN that can be used to generate synthetic data samples that are like a given dataset. The algorithm is based on the idea of using a GAN to learn the distribution of a dataset and then using that learned distribution to generate new data samples. This allows the CTGAN to generate new data samples that are similar to the original dataset.

The CTGAN algorithm consists of two parts: a generator and a discriminator. The generator is resposible for generating new data samples, while the discriminator is responsible for determining whether a given data sample is real or fake [6]. The generator and discriminator are trained jointly in an adversarial fashion. The generator tries to fool the discriminator by generating fake data samples, while the discriminator tries to identify the fake data samples. As the training progresses, the generator gets better at generating fake data samples that are close to the real data, and the discriminator gets better at identifying the fake data. Once the training is complete, the generator can be used to generate new data samples. The generated data samples will be close to the real data but will not be identical to it.

The application areas of CTGAN are manifold. It can be used for image segmentation, object detection, 3D printing, and many more. In the medical field, CTGAN can be used for image reconstruction, image registration, and image segmentation. It can also be used for 3D printing of medical images for surgical planning and training. In the retail industry, CTGAN can be used for product recognition and identification. It can also be used for the 3D printing of customized products. In the security industry, CTGAN can be used for face recognition and identification. It can also be used for the 3D printing of security-related objects.

Now let's give information about SurfelGAN.

SurfelGAN

SurfelGAN is a GAN that uses a combination of deep learning and reinforcement learning to generate realistic images of surfers [7]. The network is composed of two parts: a generator and a discriminator. The generator is responsible for creating the images, while the discriminator is responsible for distinguishing between generated images and real images [8]. The two parts are trained simultaneously in a feedback loop, with the generator trying to fool the discriminator and the discriminator trying to differentiate between real and generated images. This process results in a network that can generate realistic images of surfers.

SurfelGAN is a GAN designed for generating realistic 3D surfel (surface + volume) models. SurfelGAN can not only generate high-quality 3D surfel models but also faithfully reproduce the textures of the input data. One of the key advantages of SurfelGAN is its ability to generate realistic 3D models with a high degree of detail. In comparison, traditional 3D modeling methods often produce models with a lower level of detail. This is particularly noticeable when the models are viewed from a distance. Another advantage of SurfelGAN is its ability to faithfully reproduce the textures of the input data. This is particularly important for applications such as 3D printing and VR, where the textures of the models play a critical role in the overall experience.

SurfelGAN is a powerful GAN that can be used to generate realistic images of waves, waterfalls, rivers, lakes, and the ocean. It can also be used to generate realistic images of sunsets, clouds, and the night sky. Additionally, SurgelGAN can be used to generate realistic images of the stars, the Milky Way, and the planets. Finally, SurfelGAN can be used to generate realistic images of comets, asteroids, meteor showers, the Northern Lights, and the Southern Lights.

Now, let's give brief information about Cycle GANs.

Cycle GANs

Cycle GANs are a type of GANs that uses a recurrent neural network (RNN) to generate successive images in a sequence. The RNN is initialized with a random image, and then it predicts the next image in the sequence by considering both the current image and the previous image. This allows the generator to produce images that are more coherent and consistent with each other. In a standard GAN setup, the generator and discriminator are trained in alternation, with the generator trying to produce samples that fool the discriminator and the discriminator trying to distinguish between generated samples and real data. However, in a cycle GAN setup, the generator and discriminator are trained together, so that the generator can learn to produce images that are more consistent with the discriminator's expectations. This can result in better-quality generated images.

The Cycle GAN is a machine learning model that can be used to generate new images from a given set of images or to convert a given image into a different style. It consists of two parts: a *discriminator network*, which is used to distinguish between the original images and the generated images, and a *generator network*, which is used to generate the new images. The Cycle GAN is trained using a reinforcement learning algorithm, which allows it to learn how to generate new images that are like the originals.

There are many potential for CycleGAN applications for Cycle GANs. One example is using Cycle GANs to generate new images of people. This could be used to create realistic 3D images of people or to create new images of people that look very different from what is currently available. Another potential application is using Cycle GANs to generate new text. This could be used to create realistic dialogue for video games or movies or to generate new articles or stories. In addition, Cycle GANs could be used to improve the accuracy of machine learning algorithms.

Now, let's give brief information about SinGAN-Seg.

SinGAN-Seg

SinGAN-Seg is a new approach to generating synthetic training data for medical image segmentation. This approach uses a GAN to learn the distribution of pixel values in a given image, and then generate new images that match that distribution. This allows for the creation of large, realistic training datasets that can be used to train machine learning models for segmentation tasks. Previous approaches to generating synthetic training data have focused on creating images that are similar to a given target image. However, these approaches often fail to capture the true distribution of pixel values in the target image, resulting in training data that is not realistic. SinGAN-Seg overcomes this limitation by learning the distribution of pixel values from the target image and then generating new images that match that distribution. This approach is effective at generating realistic synthetic training data for medical image segmentation tasks. In one example, the authors generated a synthetic dataset of brain MRI images and used it to train a segmentation model. The model trained on the synthetic data was able to achieve similar performance to a model trained on a real dataset. SinGAN-Seg is a promising new approach to generating synthetic training data for medical image segmentation. This approach has the potenial to create large, realistic training datasets that can be used to train machine learning models for segmentation tasks.

SinGAN is a GAN proposed by Taesung Park et al. in the paper "Unpaired Image-to-Image Translation using Cycle-Consistent Adversarial Networks" [9]. It is used for unpaired image-to-image translation, meaning that it can learn to translate from one domain to another without having pairs of corresponding images in both domains. For example, it can learn to translate pictures of summer into pictures of winter, or pictures of dogs into pictures of cats, without having any pairs of summer-winter or dog-cat images. SinGAN has been used for a variety of applications, such as photo-realistic image generation, text-to-image synthesis, and style transfer.

Now, let's give brief information about MedGAN.

MedGAN

The MedGAN is a GAN designed to generate synthetic medical images. The network is trained on a dataset of real medical images and learns to generate new images that are realistic enough to fool a discriminator network. The MedGAN has potential applications in data augmentation, creating new training data for machine learning algorithms, or generating new images for research purposes.

The MEdGAN consists of two networks: a generator and disciriminator. The generator network is responsible for creating new images, while the discriminator network evaluates the realism of the generated images. The two networks are trained together in a competition: the generator tries to fool the discriminator by generating realistic images, while the discriminator tries to correctly classify the real and generated images.

The MedGAN has been shown to generate realistic images of various medical modalities, including X-rays, MRI scans, and histology slides. The generated images can be used to augment training data for machine learning algorithms or to generate new images for research purposes. The MedGAN is an important tool for medical image synthesis and has potential applications in many areas of medicine. It can be used to generate images of diseases or injuries, which can be used to train medical professionals. It can also be used to generate images of healthy tissue, which can be used to help researchers understand how diseases develop and how they can be treated.

MedGAN can be used for a variety of purposes, including generating realistic images for training data in computer vision, improving the performance of image recognition algorithms, creating new images from scratch, editing images to change their content, enhancing images for better visual quality, removing image artifacts and noise, compressing images for storage or transmission, generating 3D images from 2D ones, creating artistic images, and manipulating images for fun or pranks. In addition, MedGAN can be used to create images that are impossible or very difficult to create with traditional image generation methods. For example, MedGAN can be used to generate images of people with a third arm or to generate images of animals with human features.

Now, let's give brief information about DCGAN.

DCGAN

A *Deep Donvolutional Generative Adversarial Network* (DCGAN) is a type of GAN used to generate new images from a given training dataset. A GAN's consist of two networks, a generator network, and a discriminator network, that compete in a game-like fashion. The generator creates new images that are then fed into the discriminator, which attempts to classify them as real or fake. The goal of the generator is to generate images that fool the discriminator, while the goal of the discriminator is to correctly classify the images as real or fake.

DCGAN are a type of GAN that uses uses deep convolutional layers in both the generator and discriminator. This makes them well-suited for generating images, as convolutional layers can extract features from images. DCGANs have been used to generate images of faces, animals, and objects, and can even be used to create new images from a given training dataset.

To train a DCGAN, the generator and discriminator are first initialized with random weights. The generator creates new images, which are then fed into the discriminator. The discriminator attempts to classify the images as real or fake, and the generator tries to fool the discriminator. The game continues until the discriminator can no longer tell the difference between the real and fake images, at which point the DCGAN has converged.

Once trained, the DCGAN can be used to generate new images. To do this, the generator is given a noise vector, which is then used to generate a new image. The noise vector can be thought of as a seed that will determine the overall structure of the generated image. By changing the noise vector, different images can be generated.

DCGANs have shown to be successful in generating high-quality images. They are also easy to train, as they do not require a lot of data. However, DCGANs are not without their drawbacks. One issue is that the generator can sometimes generate images that are too realistic, which can be unsettling. Another issue is that DCGANs can be difficult to control, as because the noise vector can sometimes generate images that are not what was intended.

Overall, DCGANs are a powerful tool for generating new images. They are easy to train and can generate high-quality images. However, they are not without their drawbacks, and further research is needed to improve them.

Now let's give brief information about WGAN.

WGAN

WGAN, abbreviated as "Wasserstein GAN", is a GAN model. Gan models are artificial neural network model that are used to create new images using two competing neural networks. WGAN is a subset of GAN models, and GAN models have several features. One of the these features is that it uses a distance called the "*Wasserstein distance*" that GAN models use. This distance allows GAN models to produce more accurate images. Also, WGAN models require less training than GAN models and can produce images with less noise.

The WGAN is important because it allows for the creation of new, high-quality data sets that can be used to train machine learning models. This is especially important for data-intensive applications such as computer vision and natural language processing. The WGAN is also important because it can be used to improve the performance of existing machine learning models. For example, the WGAN can be used to fine-tune the weights of a neural network. Finally, the WGAN is important because it can be used to create synthetic data set. This is useful for data augmentation and for creating training data for machine learning models.

WGAN is a machine learning algorithm that is especially used for the classification of image data. It can be used in real-time face recognition and imaging systems, as well as for operations such as rebuilding, scaling, and transforming images. Additionally, WGAN can be used for video creation, encoding, and decoding, and has applications in fields such as data mining and computer graphics. Furthermore, WGAN can be used in machine learning and artificial intelligence applications and quantum computers, and the creation of digital environments. Finally, WGAN can be used for audio and video encoding, decoding, and processing, and has potential applications in computer networks and distributed systems.

Now, let's give brief information about SeqGAN.

SeqGAN

The SeqGAN is a type of generative adversarial network (GAN) designed to generate realistic sequences, such as text, using a recurrent neural network (RNN). In the paper "SeqGAN: Sequence Generative Adversarial Nets with Policy Gradient" [10]. The seqGAN is an extension of the GAN framework to sequential data. It is designed to address the issue of mode collapse, which is a common problem in GAN training. Mode collapse occurs when the generator only produces a limited number of different outputs, rather than a variety of outputs.

The seqGAN overcomes mode collapse by using a reinforcement learning technique called the *policy gradient*. The policy gradient is a method of optimizing a stochastic policy by gradient descent. In seqGAN, the policy gradient is used to update the generator's parameters so that it produces a variey of outputs.

The seqGAN is effective at generating realistic text. In the paper "Generating Wikipedia by Summarizing Long Sequences" [11], the seqGAN was used to generate text by summarizing long sequences of text. The seqGAN was also used in the paper "Adversarial Feature Matching for Text Generation" to generate text that is indistinguishable from the real text [12].

The seqGAN is composed of two components: a generator and a discriminator. The generator is responsible for generating fake data, while the discriminator tries to classify the data as either real or fake. The generator is trained by maximizing the probability of the discriminator making a mistake. In other words, the generator is trying to fool the discriminator. The discriminator, on the other hand, is trained to correctly classify the data.

The main advantage of the SeqGAN over other GAN models is that it can generate high-quality synthetic data. In addition, the seqGAN can capture the long-term dependencies in sequential data. The seqGAN has been used to generate synthetic data for various applications, such as text generation, music generation, video generation, sentiment analysis, dialogue generation, text summarization, article spinning, text classification, natural language processing, question answering, machine translation, and Information retrieval.

Now, let's give brief information about Conditional GAN.

Conditional GAN

A conditional GAN (cGAN) is a type of GAN used to generate images. They are like GANs but differ in that they use a conditional input to generate images. This means that they can be trained to generate images based on a given input, such as a photograph. This makes them useful for tasks such as image completion and image generation.

It can be used to generate images from a given input. The input can be anything from a noise vector to an image. The cGAN can be used to generate images of faces from a given input of noise vectors. The cGAN can also be used to generate images of faces from an input of images of faces. One advantage of using a cGAN is that the model can learn to generate highly realistic images. Another advantage of using a cGAN is that the model.

The cGAN is a powerful tool for image generation. The cGAN can be used to generate images that are not possible to generate with traditional methods. For example, the cGAN can be used to generate images of faces that are not possible to generate with traditionally.

The cGAN consists of two parts: the generator and the discriminator. The generator is responsible for generating images, and the discriminator is responsible for discriminating between real and generated images. The generator is trained to generate images that are as realistic as possible, while the discriminator is trained to discriminate between real and generated images. The training process is an adversarial process, where the generator and discriminator are competing against each other [13]. The cGAN can be used to generate images that are conditioned on some input, such as a class label. For example, a cGAN could be used to generate images of faces that are conditioned on the identity of the person in the image. This can be useful for generating images in applications where the input is known, such as image synthesis and image completion.

Finally, let's give a piece of brief information about BigGAN.

BigGAN

BigGAN is a deep learning model that can generate realistic images from scratch. It was first introduced in a paper by Google AI researchers in 2018 [14]. BigGAN is a large-scale generative model that is trained on a dataset of images to generate new images. The model is designed to be able to generate high-quality images and can do so by using several techniques, such as using many parameters and making use of a hierarchical latent space.

BigGAN is a model for generating high-resolution images. BigGAN is based on a GAN or generative adversarial network. A GAN consists of two parts: a generator and a discriminator. The generator creates images, and the discriminator tries to distinguish between real and generated images [15]. The two parts are trained together, and the generator gets better at creating images that the discriminator can't tell apart from real images.

GANs are a powerful tool for generative modeling but training them can be difficult. BigGAN is a new method for training GANs that promises to make training easier and more effective. BigGAN is based on a technique called *self-attention*, which allows the network to focus on important details in the data. This makes it possible to train a GAN to generate high-quality images with less data and less training time. BigGAN has been used to generate realistic images of faces, animals, and scenes. This could potentially be used to create realistic images for use in computer vision applications. Self-attention is a technique that allows a network to focus on important details in the data. This makes it possible to train a GAN to generate high-quality images with less data and less training time.

BigGAN is an extension of the GAN model that can generate images of much higher resolution than previous models. It does this by using a new technique called "*projection discriminators*". Projection discriminators can learn much more complex relationships between images and their labels than previous discriminator models. This allows BigGAN to generate images that are much more realistic and diverse than previous models.

BigGAN uses several techniques to improve the quality of the generated images, including a hierarchical latent space, a truncation trick, and a new GAN loss function. The hierarchical latent space allows the network to generate images that are more varied and realistic than those generated by previous GANs. The truncation trick allows the network to generate images that are closer to the training data, while the new GAN loss function encourages the network to generate images that are both realistic and diverse.

Summary

In this chapter, you learned about Generative Adversarial Networks or GANs. You learned about the different typesof GANs, including CTGANs, SurfelGANs, Cycle GANs, SinGANs, medGANs,DCGANs, WGANs, seqGANs, and Conditional GANs.You also learned about the BigGAN and how it can be used to generate realistic images.

Next, we'll being delving into Synthetic Data Generation with R.

References

[1]. I. J. Goodfellow et al., "Generative Adversarial Nets," Jun. 2014. Accessed: Apr. 13, 2022. [Online]. Available: `http://www.github.com/goodfeli/adversarial`.

[2]. A. Rosolia and J. Osterrieder, "Analyzing Deep Generated Financial Time Series for Various Asset Classes," SSRN Electronic Journal, 2021, doi: 10.2139/ssrn.3898792.

[3]. X. Chen et al., "Enhanced dielectric properties due to space charge-induced interfacial polarization in multilayer polymer films," J. Mater. Chem. C, vol. 5, no. 39, pp. 10417–10426, 2017, doi: 10.1039/C7TC03653A.

[4]. M. Zhang, N. A. Deskins, G. Zhang, R. T. Cygan, and M. Tao, "Modeling the Polymerization Process for Geopolymer Synthesis through Reactive Molecular Dynamics Simulations," The Journal of Physical Chemistry C, vol. 122, no. 12, pp. 6760–6773, Mar. 2018, doi: 10.1021/acs.jpcc.8b00697.

[5]. D. Ravi, S. B. Blumberg, S. Ingala, F. Barkhof, D. C. Alexander, and N. P. Oxtoby, "Degenerative adversarial neuroimage nets for brain scan simulations: Application in ageing and dementia," Medical Image Analysis, vol. 75, p. 102257, Jan. 2022, doi: 10.1016/j.media.2021.102257.

[6]. E. Wang, L. Xue, Y. Li, Z. Zhang, and X. Hou, "3DMGNet: 3D Model Generation Network Based on Multi-Modal Data Constraints and Multi-Level Feature Fusion," Sensors, vol. 20, no. 17, p. 4875, Aug. 2020, doi: 10.3390/s20174875.

[7]. Z. Yang et al., "SurfelGAN: Synthesizing Realistic Sensor Data for Autonomous Driving," in 2020 IEEE/CVF Conference on Computer Vision and Pattern Recognition (CVPR), Jun. 2020, pp. 11115–11124, doi: 10.1109/CVPR42600.2020.01113.

[8]. X. Dai, X. Yuan, and X. Wei, "Data augmentation for thermal infrared object detection with cascade pyramid generative adversarial network," Applied Intelligence, vol. 52, no. 1, pp. 967–981, Jan. 2022, doi: 10.1007/S10489-021-02445-9.

[9]. J. -Y. Zhu, T. Park, P. Isola, and A. A. Efros, "Unpaired Image-To-Image Translation Using Cycle-Consistent Adversarial Networks," in Proceedings of the IEEE International Conference on Computer Vision (ICCV), 2018, pp. 2223–2232. Accessed: Apr. 19, 2022. [Online]. Available: `https://ieeexplore.ieee.org/stamp/stamp.jsp?tp=&arnumber=8237506`.

[10]. L. Yu, W. Zhang, J. Wang, and Y. Yu, "SeqGAN: Sequence Generative Adversarial Nets with Policy Gradient," 2017. Accessed: Apr. 20, 2022. [Online]. Available: `www.aaai.org`

[11]. P. J. Liu et al., "Generating Wikipedia by Summarizing Long Sequences," 2018. Accessed: Apr. 20, 2022. [Online]. Available: `https://en.wikipedia.org/wiki/Wikipedia:Manual_of_Style`

[12]. Y. Zhang et al., "Adversarial Feature Matching for Text Generation," in Proceedings of the 34th International Conference on Machine Learning, PMLR, 2017, pp. 4006–4015.

[13]. P. Salehi, A. Chalechale, and M. Taghizadeh, "Generative Adversarial Networks (GANs): An Overview of Theoretical Model, Evaluation Metrics, and Recent Developments," 2020.

[14]. S. -W. Park, J. -S. Ko, J. -H. Huh, and J. -C. Kim, "Review on Generative Adversarial Networks: Focusing on Computer Vision and Its Applications," Electronics (Basel), vol. 10, no. 10, p. 1216, May 2021, doi: 10.3390/electronics10101216.

[15]. M. Castelli, L. Manzoni, T. Espindola, A. Popovič, and A. de Lorenzo, "Generative adversarial networks for generating synthetic features for Wi-Fi signal quality," PLOS ONE, vol. 16, no. 11, p. e0260308, Nov. 2021, doi: 10.1371/journal.pone.0260308.

CHAPTER 4

Synthetic Data Generation with R

In this chapter, we will explore how to generate synthetic data using the R programming language. We'll start by looking at some of the basic functions used to generate synthetic data. Next, we'll look at how to construct a value vector from a known univariate distribution. Next, we'll look at how to construct a vector from a multi-level categorical variable. We will look at how to build a neural network in R using the nnet package. We will give an example of image augmentation using a torch package. Next, we'll explore how to generate synthetic data in R using the mice conjurer and synthpop package. Finally, the topic of copulas and R practices related to copula, normal copula, and Gaussian copula will be covered.

Basic Functions Used in Generating Synthetic Data

A synthetic data generation method is an approach to creating new, artificial data that resembles real data in some way. There are many ways to generate synthetic data, but all methods share the same goal: to create data that can be used to train machine learning models without the need for real data.

Basic functions are used in synthetic data generation for a variety of reasons. First, basic functions allow for the construction of new variables that are not present in the original data set. This is important because it allows for the creation of variables that may be important for the analysis but were not originally measured. Second, basic functions can be used to transform variables in the original data set. This is important because it allows for the creation of new variables that are more appropriate for the analysis. Finally, basic functions can be used to create missing values in the data set. This is important because it allows for the creation of a more realistic data set that includes missing values.

N. Gürsakal et al., *Synthetic Data for Deep Learning*, https://doi.org/10.1007/978-1-4842-8587-9_4

There are functions in the R programming language that can be used to create synthetic datasets. These datasets can be used to test code used in R programming.

The codes used in this example are from `https://projets.pasteur.fr/projects/rap-r/wiki/Synthetic_Data_Generation`.

```
# Concatenate into a vector
> a=c(3,4,6,2,1)
> b=c(6,3,2,8,9, a)
> b
 [1] 6 3 2 8 9 3 4 6 2 1

# Create an increasing array from 7 to 14 with an interval of 1.4
> seq(from=7, to=14,by=1.4)
[1]  7.0  8.4  9.8 11.2 12.6 14.0

# Create a sequence that repeats 7 5 times
> rep(7,times=5)
[1] 7 7 7 7 7

# Define a vector y that can take a value between 4 and 20.
> y<-4:20
> y
 [1]  4  5  6  7  8  9 10 11 12 13 14 15 16 17 18 19 20

# A random permutation
> sample(y)
 [1] 16 13 17 14 10  8  5 11 20  4  9 12 15 19 18  6  7

# Bootstrap resampling
> sample(y,replace=TRUE)
 [1] 13 12 20 16 18  4 17  9 10 10 18  9 17 20 13 17 15

# Bernoulli trials
> sample(c(3,7),10,replace=TRUE)
 [1] 7 3 7 3 3 7 3 3 3 7
```

Creating a Value Vector from a Known Univariate Distribution

To obtain the same result in random operations, the "set.seed()" function is written in front of the commands.

```
> set.seed(12)
# Select sample number as 15
> n=15
# Uniform distribution btw 5 and 30
> runif(n, min=5, max=30)
[1] 15.983358 16.440179 18.517689 21.641996  7.817473 10.459179
[7] 24.695909  7.446326 22.745762 10.445576 11.698590 17.619199
[13]  9.714673 15.985733 21.745482

# A Gaussian distribution has a mean of 3 and a standard deviation of 1.5.
> rnorm(n, mean=3, sd=1.5)
[1] 4.0109717 6.1080537 2.1884570 1.3942618 2.4413149
[6] 2.2722880 3.4121763 2.2807312 4.1971580 1.4933232
[11] 3.1574763 1.2660107 3.8672019 0.6065615 2.5372445
# The parameter lambda is used to determine the shape of the Poisson
distribution.
> rpois(n, lambda=3)
[1] 4 5 5 6 4 4 3 4 4 3 4 2 7 1 0
# The proportional exponential distribution can be used to model the
waiting time for an event occurring at a continuous rate.
> rexp(n, rate=2)
 [1] 0.2498384 0.5392200 0.4350614 0.8315090 0.7619429 0.2520267 0.5561264
 [8] 0.2639337 0.4741497 0.2147715 0.1082590 0.6139462 0.4781476 0.2432179
[15] 0.8234695

# The binomial distribution with size and prob is a way of looking at
data that is collected in a certain way. This distribution is used when
there are a certain number of trials, and when each trial has two possible
outcomes, which are either success or failure.
> rbinom(n, size=5, prob=0.5)
[1] 4 4 3 4 3 2 4 3 3 2 3 2 4 2 2
```

```
# lognormal distribution
> rlnorm(n, meanlog=2, sdlog=1.5)
[1]  2.855488  1.097922  4.154011 16.040772  5.657868  7.436401  1.092983
[8]  5.456647 42.380944  7.134419 28.381432  5.668436 39.274035  3.277781
[15]  1.741771
```

Vector Generation from a Multi-Levels Categorical Variable

```
# Generating a random sequence from a four-level categorical variable.
> sample(c("G-1","G-2","G-3","G-4"),8,replace=TRUE)
[1] "G-4" "G-3" "G-3" "G-3" "G-4" "G-3" "G-4" "G-1"
# A five-level categorical variable can be used to generate a random
sequence.
> sample(c("G-1","G-2","G-3","G-4","G-5"),10,replace=TRUE)
 [1] "G-5" "G-1" "G-1" "G-5" "G-5" "G-4" "G-5" "G-3" "G-2" "G-5"
```

Multivariate

```
# This exercise will generate a data.frame from 4 different samples, each
with 5 different variables from the same distribution.
> data.frame(indv=factor(paste("S-", 1:4, sep = "")), matrix(rnorm(4*5, 4,
2), ncol = 5))
  indv       X1       X2       X3       X4       X5
1  S-1 3.221146 2.000166 4.796590 3.033753 2.877777
2  S-2 5.373274 2.557023 3.372633 5.074982 4.614122
3  S-3 4.380217 4.593711 6.325639 4.058779 3.620474
4  S-4 1.317684 6.412876 3.259470 6.295884 5.125372

# This code will generate a data.frame with 3 independent variables from 4
different distributions.
> data.frame(indv=factor(paste("S-", 1:4, sep = "")), W1=rnorm(4,
mean=3,sd=2), W2=rnorm(4,mean=5, sd=3), W3=rpois(4,lambda=5))
indv       W1       W2 W3
1  S-1 4.123975 2.045277  4
2  S-2 4.740534 6.300097  5
```

```
3  S-3 1.470572 6.263931  3
4  S-4 0.430045 2.160229  5

# This data.frame has 1 categorical variable (2 levels) and 3 independent
continuous variables.
> data.frame(indv=factor(paste("S-", 1:8, sep = "")), W1=rnorm(8,
mean=1,sd=2), W2=rnorm(8,mean=10, sd=4), W3=rnorm(8,mean=5,sd=3),
Animal=sample(c("Cat","Dog"),8, replace=TRUE))
indv         W1          W2        W3    Animal
1  S-1  1.8277723  4.5727621 5.261963    Dog
2  S-2 -1.3343894 14.9566491 4.743537    Dog
3  S-3  3.3592723 15.2376011 5.333706    Dog
4  S-4 -2.8445692 12.1955695 5.677764    Cat
5  S-5  1.8944436 18.6326683 3.860072    Dog
6  S-6  1.2227463  0.4451184 5.478651    Cat
7  S-7 -3.7531096  6.1432943 7.847106    Cat
8  S-8 -0.6455443  6.1451687 3.093108    Dog
```

Multivariate (with correlation)

The MASS package can be used to generates samples from a multivariate normal distribution. This can be useful for simulation studies or for estimating parameters. For example, we can create 3 random samples from a multivariate normal distribution using the MASS package. This can be useful for testing statistical methods or for exploring the data. The MASS package can also be used to generate samples from other distributions, which can be useful for different purposes.

```
# Load "MASS", "psych" and "rgl" packages.
> library(MASS)
> library(psych)
> library(rgl)
> set.seed(50)
> m <- 3
> n <- 1000
> sigma <- matrix(c(1, -0.4, 0.3, -0.8, 1, 0.6, -0.7, 0.2, 1), nrow=3)
> X <- mvrnorm(n, mu=rep(0, m),Sigma=sigma,empirical=T)
```

```
> colnames(X) <- paste0("X", 1:m)
> cor(X,method='spearman')
           X1         X2        X3
X1  1.0000000 -0.3859910 0.2888176
X2 -0.3859910  1.0000000 0.5766619
X3  0.2888176  0.5766619 1.0000000
```

```
# Compare variables
> pairs.panels(X)
```

The output is shown in Figure 4-1.

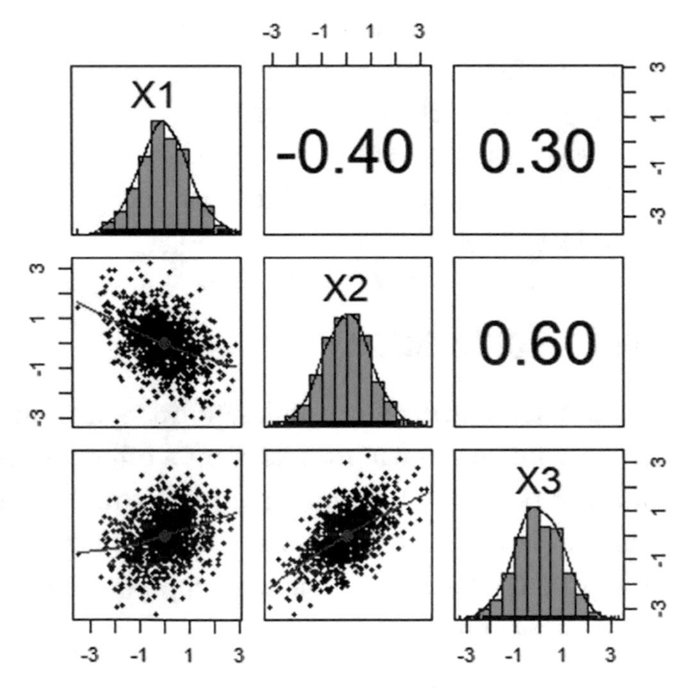

***Figure 4-1.** Marginal normal distribution of variables*

```
# Normalize the variables
>w <- pnorm(X)
# Compare normalized variables
>pairs.panels(w)
```

The output is shown in Figure 4-2.

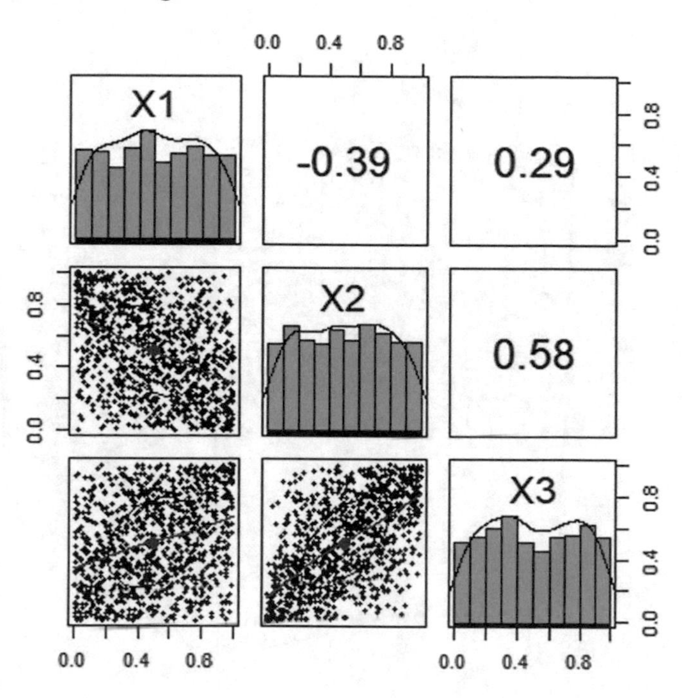

***Figure 4-2.** Distributions of normalized variables*

Here the "rgl" package will be used to visualize the data in three-dimensional space. The "rgl" package is a graphics device system for R. It provides medium to high-level graphics functions for three-dimensional data visualization, including functions like point clouds, mesh surfaces, volumetric representations, and geometric primitives.

```
# Draw the representation of variables in 3D space using the "rgl" package.
>plot3d(w[,1],w[,2],w[,3],pch=30,col='black')
```

The output is shown in Figure 4-3.

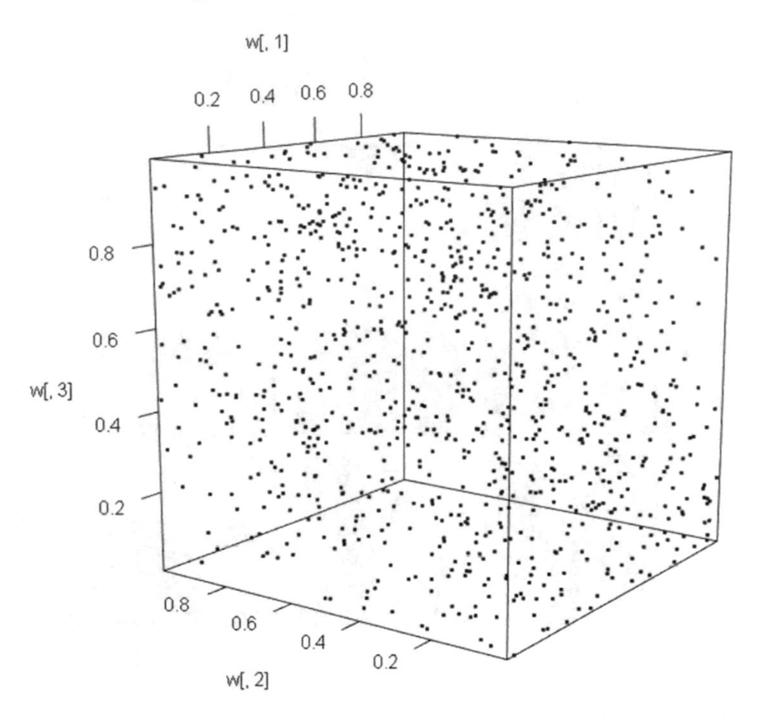

Figure 4-3. *Representation of variables in 3D space*

```
# Create variables u1, u2 and u3
> u1 <- qt(w[,1],df=5)
> u2<-qgamma(w[,2],shape=2,scale=1)
> u3 <- qbeta(w[,3],3,3)

# Draw a graph for the variables u1, u2, and u3 in a 3-dimensional space
>plot3d(u1,u2,u3,pch=20,col='blue')
```

The output is shown in Figure 4-4.

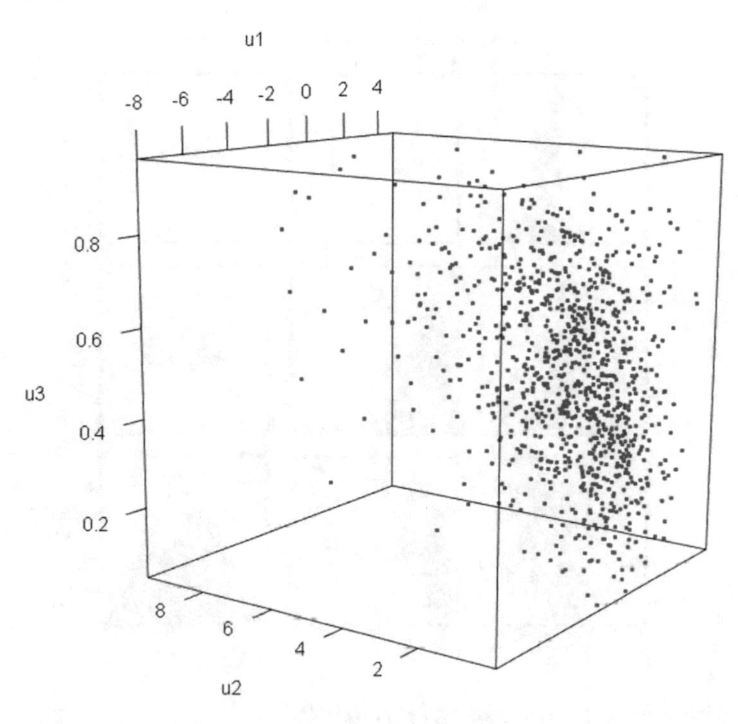

Figure 4-4. *3D drawing of simulated data*

```
# Creates a data frame using the simulated variables u1, u2, and u3.
>data.frame<-cbind(u1,u2,u3)

# Print Spearman correlation matrix of variables
>cor(data.frame,meth='spearman')
          u1         u2        u3
u1  1.0000000 -0.3859910 0.2888176
u2 -0.3859910  1.0000000 0.5766619
u3  0.2888176  0.5766619 1.0000000
# Compare simulated variables
>pairs.panels(data.frame)
```

The output is shown in Figure 4-5.

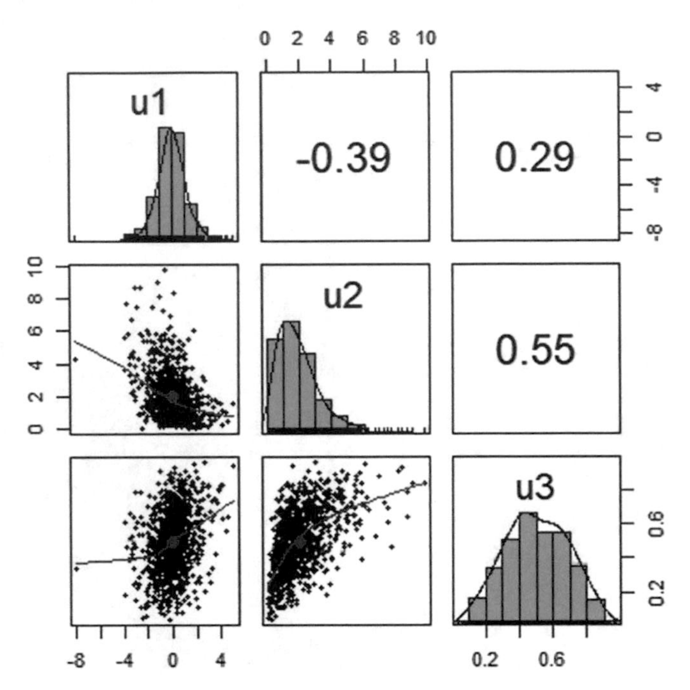

Figure 4-5. *Marginal distributions of variables*

Generating an Artificial Neural Network Using Package "nnet" in R

Neural networks are powerful tools for modeling complex patterns in data. The main difference between these tools and other machine learning tools is that they consist of many interconnected neurons that can learn to recognize input patterns. Artificial neural networks are widely used in many different applications, including pattern recognition, classification, and prediction. For example, they can be used in security cameras; in authentication processes, to determine who a person is; They can be used to authenticate at a bank ATM. The neural network can also work with many different types of sensors and process many different types of input data.

In this example, you will learn how to use the "nnet" package to create and train a simple neural network. The "nnet" package in R provides users with a variety of tools to create and train neural networks. The "nnet" package includes functions for creating and training neural networks, as well as for evaluating the performance of a neural network model. The package also includes functions for predicting the output of a neural network.

The data used in this example are from https://archive.ics.uci.edu/ml/datasets/Glass+Identification and the codes used in this example are from "https://rpubs.com/khunter/cares_neuralnets".

```
# Read the dataset from the file "C:/Users/........./glass.csv" and store
it in a variable named "glass"
> glass = read.csv("C:/Users/........./glass.csv")
# Generate an ID number for each column in the dataset
>glass$id = as.character(seq(1, nrow(glass)))
# Print the first six rows in the dataset
>head(glass)
Id.number refractive.index Sodium Magnesium Aluminum Silicon Potassium
1             1.52    13.6       4.49      1.1      71.8       0.06
2             1.52    13.9       3.6       1.36     72.7       0.48
3             1.52    13.5       3.55      1.54     73.0       0.39
4             1.52    13.2       3.69      1.29     72.6       0.57
5             1.52    13.3       3.62      1.24     73.1       0.55
6             1.52    12.8       3.61      1.62     73.0       0.64
# … with 5 more variables: Calcium <dbl>, Barium <dbl>, Iron <dbl>,

# Modify the glass variable to create new variables for each level of
the Type.of.glass variable. The new variables are called label1, label2,
label3, label4, label5, label6, and label7. The Type.of.glass variable is
then converted into a factor.
>glass.label = mutate( glass, label1 = Type.of.glass== '1', label2 =Type.
of.glass== '2', label3 = Type.of.glass== '3', label4 =Type.of.glass==
'4',label5 = Type.of.glass== '5', label6 =Type.of.glass== '6',label7 =
Type.of.glass== '7',Type.of.glass = factor(Type.of.glass) )

# Convert "glass.label" to a class vector.
>sapply(glass.label, class)
Id.number refractive.index            Sodium         Magnesium
       "numeric"         "numeric"         "numeric"         "numeric"
        Aluminum           Silicon         Potassium           Calcium
       "numeric"         "numeric"         "numeric"         "numeric"
          Barium              Iron    Type.of.glass                id
       "numeric"         "numeric"          "factor"       "character"
```

```
     label1          label2          label3          label4
  "logical"       "logical"       "logical"       "logical"
     label5          label6          label7
  "logical"       "logical"       "logical"
> feature.names = colnames(glass)[!(colnames(glass) %in% c('id',
'Id.number' ,'Type.of.glass', 'label1', 'label2','label3',
'label4','label5', 'label6','label7'))]
# Test whether each variable in the dataset "glass.label" is a
numeric value.
>numeric = sapply(glass.label, is.numeric)
>numeric
Id.number refractive.index          Sodium       Magnesium
       TRUE            TRUE            TRUE            TRUE
   Aluminum         Silicon       Potassium         Calcium
       TRUE            TRUE            TRUE            TRUE
     Barium            Iron   Type.of.glass              id
       TRUE            TRUE           FALSE           FALSE
     label1          label2          label3          label4
      FALSE           FALSE           FALSE           FALSE
     label5          label6          label7
      FALSE           FALSE           FALSE

>glass.scaled = glass.label
# Scale the dataset
>glass.scaled[ ,numeric]= sapply(glass.label[,numeric], scale)
# Print the first six lines of the scaled dataset.
>head(glass.scaled)
  Id.number refractive.index     Sodium Magnesium   Aluminum     Silicon
Potassium
1 -1.719943        0.8708258  0.2842867 1.2517037 -0.6908222 -1.12444556
-0.67013422
2 -1.703794       -0.2487502  0.5904328 0.6346799 -0.1700615  0.10207972
-0.02615193
3 -1.687644       -0.7196308  0.1495824 0.6000157  0.1904651  0.43776033
-0.16414813
```

```
4 -1.671494       -0.2322859 -0.2422846 0.6970756 -0.3102663 -0.05284979
0.11184428
5 -1.655344       -0.3113148 -0.1688095 0.6485456 -0.4104126  0.55395746
0.08117845
6 -1.639195       -0.7920739 -0.7566101 0.6416128  0.3506992  0.41193874
0.21917466
     Calcium     Barium       Iron Type.of.glass id label1 label2 label3
label4
1 -0.1454254 -0.3520514 -0.5850791             1  1   TRUE  FALSE  FALSE
FALSE
2 -0.7918771 -0.3520514 -0.5850791             1  2   TRUE  FALSE  FALSE
FALSE
3 -0.8270103 -0.3520514 -0.5850791             1  3   TRUE  FALSE  FALSE
FALSE
4 -0.5178378 -0.3520514 -0.5850791             1  4   TRUE  FALSE  FALSE
FALSE
5 -0.6232375 -0.3520514 -0.5850791             1  5   TRUE  FALSE  FALSE
FALSE
6 -0.6232375 -0.3520514  2.0832652             1  6   TRUE  FALSE  FALSE
FALSE
  label5 label6 label7
1  FALSE  FALSE  FALSE
2  FALSE  FALSE  FALSE
3  FALSE  FALSE  FALSE
4  FALSE  FALSE  FALSE
5  FALSE  FALSE  FALSE
6  FALSE  FALSE  FALSE

# Generate a training dataset by randomly selecting 60 samples in the "id"
variable
>train.sample = sample(glass$id,60)

# Create a test sample based on the training sample
> test.sample = glass$id[!(glass$id %in% train.sample)]

# Scale the "train.sample" dataset
> glass.train = glass.scaled[train.sample, ]
```

```
# Scale the "test.sample" dataset
> glass.test = glass.scaled[test.sample, ]

# Create a regression model with a dependent variable "Type.of.glass"
> nnet.formula = as.formula(paste('Type.of.glass~', paste(feature.names,
collapse = ' + ')))
# Print the generated regression model
> print(nnet.formula)
Type.of.glass ~ refractive.index + Sodium + Magnesium + Aluminum +
    Silicon + Potassium + Calcium + Barium + Iron

# Upload the "nnet" and "neuralnet" libraries to the existing R session to
train neural networks.
> library(nnet)
> library(neuralnet)
> nnet.model = nnet(nnet.formula, data = glass.train, size =5)
# weights:  56
initial  value 61.056145
iter  10 value 0.375113
iter  20 value 0.002223
final  value 0.000070
converged

# Print nnet.model nnet
> nnet.model
a 9-5-1 network with 56 weights
inputs: refractive.index Sodium Magnesium Aluminum Silicon Potassium
Calcium Barium Iron
output(s): Type.of.glass
options were - entropy fitting
# Print the first six lines of the predicted pattern
> head(predict(nnet.model))
          [,1]
1 0.000000e+00
2 0.000000e+00
3 0.000000e+00
4 9.999581e-01
```

```
5 1.026017e-06
6 0.000000e+00

# Note that the left side of the formula is different for the two packages
> neuralnet_formula = paste('label1 + label2+label3 + label4+label5 +
label6+label7~', paste(feature.names, collapse = ' + '))

# Print the formula used in modeling the neural network
> print(neuralnet_formula)
[1] "label1 + label2+label3 + label4+label5 + label6+label7~ refractive.
index + Sodium + Magnesium + Aluminum + Silicon + Potassium + Calcium +
Barium + Iron"
> neuralnet.model = neuralnet(  neuralnet_formula, data = glass.
train,   hidden = c(5), linear.output = FALSE)

# Print the output results of the neural network model
> print(head(neuralnet.model$net.result[[1]]))
           [,1]        [,2]         [,3]         [,4]         [,5]
[1,] 0.99892017 0.003101246 3.197760e-06 5.021891e-07 9.660337e-07
[2,] 0.99775878 0.001284624 7.696107e-10 1.319792e-10 3.354974e-11
[3,] 0.99856318 0.004242630 4.243761e-05 2.904989e-06 1.066150e-05
[4,] 0.05082495 0.953484676 4.766107e-09 2.425787e-08 1.902702e-09
[5,] 0.97261516 0.010984899 2.072264e-10 5.220585e-11 7.683335e-12
[6,] 0.98710994 0.017597351 5.618931e-09 8.099489e-09 2.702098e-09
             [,6]         [,7]
[1,] 5.453343e-07 2.137497e-07
[2,] 2.399037e-10 4.342645e-11
[3,] 7.368009e-06 2.196349e-06
[4,] 1.618680e-08 9.323684e-10
[5,] 1.312461e-10 1.206774e-11
[6,] 6.127981e-09 4.710287e-10

# Draw the model of the neural network
>plot(neuralnet.model)
```

The output is shown in Figure 4-6.

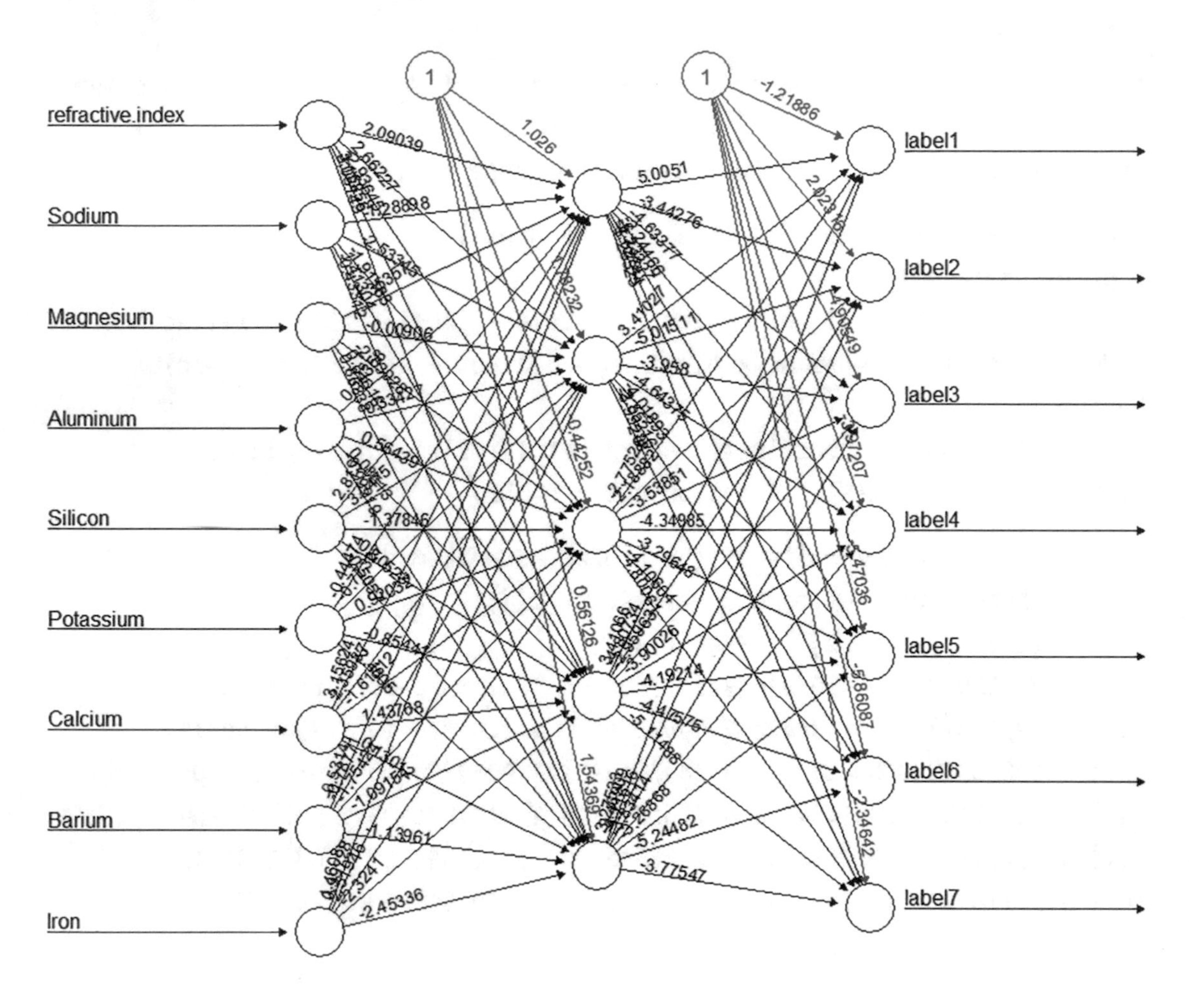

Figure 4-6. *Neural network model*

Augmented Data

The general purpose of machine learning is to automatically classify and label data or discover patterns in a dataset. Using machine learning, we can classify datasets and make predictions using the data. If we aim to classify a dataset, for example, to separate photos into cats and dogs, we need to give some of the data to the machine by labeling it. The labeling process is the process of identifying the types of data in a set of raw data, such as images, text, and photographs. From such labeled data, the computer learns whether a photo contains a cat or a tumor and can then identify classes of unlabeled data. The data labeling process is a necessary process for the computer to see and recognize images, text, and speech.

Data augmentation is a technique used in synthetic data, which is a technique used to generate new data points using a training dataset. The purpose of data augmentation is to improve the quality of the training dataset, which can help to improve the quality of the model. Data augmentation can be used to add more data points to the dataset, to improve the distribution of the data, to add more variety to the data, or to improve the data quality.

Computer vision and machine learning are ways to make images look better. Image augmentation is a technique used in these fields to make images look better. This can involve applying "geometric transformations, color space augmentations, kernel filters, mixing images, random erasing, feature space augmentation, adversarial training, GANs, neural style transfer, and meta-learning" [1]. These algorithms can be applied in different ways to produce different results.

AlexNet is a deep learning network that was developed by Alex Krizhevsky. It was one of the first networks to use data augmentation, which helped it to achieve better results. AlexNet used two types of augmentations: horizontal reflections and image translations.

All inputs in AlexNet must be of size 256x256. If the input image is not of that size, it must be converted to that size before being used for training the network.

Supervised learning is a way to get a computer to learn from data. In this approach, data is first labeled (or classified), and then the computer uses that data to learn how to do things on its own. This means that the computer is taught how to do something by being shown examples of what is wanted. Supervised learning is important to train the model using labeled data based on *"ground truth."* For example, people might ask, "Is there a bird in this photo?" when working with unlabeled data. The accuracy of the models depends on the correctness of the data. For example, the accuracy of a facial recognition model depends on the correctness of the data labeling the faces.

In natural language processing processes, important parts of the text are determined manually, and labels are given to them to create the training dataset. In this type of labeling, if sentiment analysis is to be applied, text parts, names, places, and people that determine the sentiment are labeled. Also, in such a process, certain parts of the text can be marked using a bounding box. In audio data, sounds (such as barking, chirping, bird sounds, breaking) are first converted into text and these are used as training data in structured data format.

The “ggpubr” package is an add-on for the popular “ggplot2” package for R that provides several functions for creating and manipulating publication-quality graphs. These functions include creating custom themes, adding annotations and labels, and customizing the layout and appearance of graphs. The “ggpubr” package also includes several functions for importing and exporting data to and from different formats, making it easy to create graphs that can be used in reports or presentations.

An example of how to create a text graphical object with R’s “ggpubr” package.

```
# Install ggpubr package
>install.packages("ggpubr")
# Load the "ggpubr" library
> library(ggpubr)
# Store the text to be used in the analysis in an array
> Text<-paste("The iris data set is a dataset consisting","of measurements
of the features of","150 irises. The dataset is used to","train and test
machine learning algorithms.","The training set is used to","train the
machine learning algorithm,","and the test set is used to test the accuracy
of","the machine learning algorithm.",sep="\n")

# Create a text grob
> TextGrob <- text_grob(Text, face = "italic", color = "red")

# Draw the text
> as_ggplot(TextGrob)
```

The iris data set is a dataset consisting of measurements of the features of 150 irises. The dataset is used to train and test machine learning algorithms. The training set is used to train the machine learning algorithm, and the test set is used to test the accuracy of the machine learning algorithm.

Augmentation can be accomplished by making changes to existing data or using GANs. These changes can include padding, rotating, scaling, flipping, translating, cropping, zooming, darkening, brightening, and color modification. Additionally, noise can be added, and contrast can be changed.

TF-Transformation Functions is a data augmentation method developed by Stanford AI Lab. This method is an image processing technique used to obtain data sets. TF-Transformation Functions enable datasets to be obtained by changing images and sounds. This method allows obtaining datasets by recording images from different angles and sounds at different frequencies.

How the data is transformed into augmented data with the transformation functions (listed changes to existing data) is seen in the following Figure 4-7. We can label images, pixels, or key points to allow the computer to see them, or we can create a *"bounding box"* that bounds a digital image in the process to convert the training data into a labeled form. Labeling data in this way will allow the vision model to classify images, locate objects, identify key points of the image, or identify the part of the image.

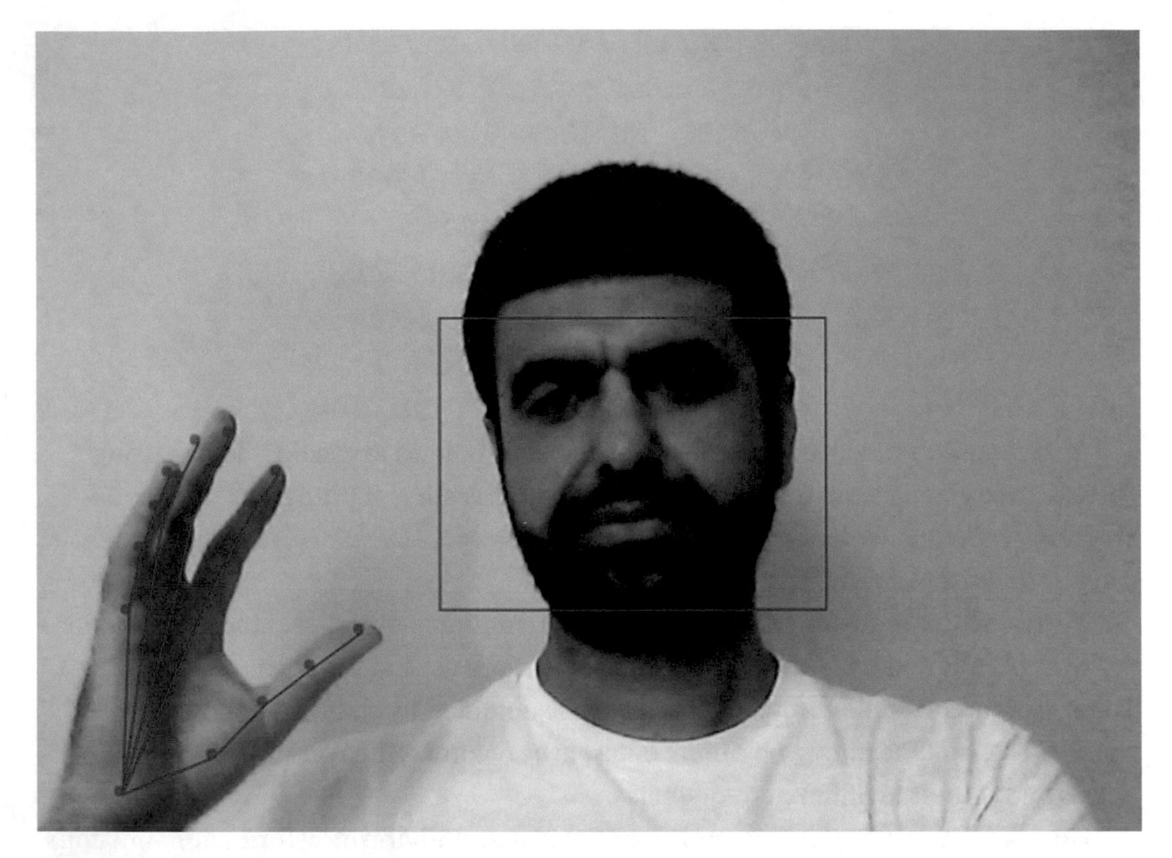

Figure 4-7. *Bounding Box*

Transformation operations such as padding, random rotating, and re-scaling, which we listed before, can be done manually, as well as automatic transformation operations can be performed on them after they are framed in the "bounding box", as seen in Figure 4-7.

There are different formats of bounding box annotations. The coordinates of the rectangular borders marking the person in the image are called *bounding boxes*. An example of bounding boxes is given in Figure 4-8.

***Figure 4-8.** An example image with a bounding box*

Data augmentation is a suitable technique to improve the efficiency of image classifiers. Augmentation is done by adding new data to existing data. For example, we can generate new data by rotating an existing photo or by decreasing or increasing its brightness. The two properties of a good tester are intra-class invariance and inter-class distinctiveness. *Intra-class invariance* means that a data point can be replaced by another data point within the same class without affecting the results of the tester. *Inter-class distinctiveness* means that a data point can be replaced by a data point from a different class without affecting the results of the tester.

GANs are used to create new data that is as similar as possible to the real data. This is done by having a generator create data that is indistinguishable from the real data, while the discriminator tries to differentiate between the real data and the generated data. This process helps to create a more accurate approximation of the "true distribution" of data.

Python packages used for data augmentation include the Keras ImageDataGenerator, Skimage, and OpenCV. Anonymizing data means removing personal information from a dataset. This is done to protect people's privacy. It is especially popular for text, a kind of structured data used in industries like finance and healthcare [2].

Anonymized and augmented data are different from synthetic data. Anonymized data is data that has been altered so that the identities of the people or organizations involved are hidden. Augmented data is data that has been altered to include extra information. Synthetic data is data that has been created artificially, using data augmentation techniques. For example, we can create a synthetic photo of cars using photos of cars.

Random erasing is a technique used in training CNNs. This technique randomly selects a rectangular region in the image and deletes its pixels with random values. This is claimed to help correct over-fitting in the model.

The Falling Things dataset is a collection of 61,500 images taken in household environments, which can be used to train and evaluate object detection and pose estimation algorithms. The dataset includes images of objects in various positions, allowing researchers to test algorithms for accurately detecting and estimating the position of objects. Additionally, the dataset includes images of objects in different depths and with different sensor modalities, allowing researchers to test algorithms for accurately estimating the depth and detecting objects in different environments.

While the data augmentation technique is applied in computer vision, Reinforcement Learning with Augmented Data (RAD) technique can be applied. In this technique, it has been demonstrated by a study that the person performing the application provides an improvement in data efficiency and generalization by learning with various views of the same input. Reinforcement learning can be more widely applied to the real world if the gap between data efficiency and generalization is bridged. This would allow for more accurate learning from data and improve the overall performance of reinforcement learning algorithms.

Image Augmentation Using Torch Package

Torch is a machine learning library that is used to do natural language processing, computer vision, and general machine learning tasks. Torch is written in Lua, which is a scripting language that is easy to learn and use. The Torch has a lot of different statistical and graphical techniques that can be used, making which makes it versatile. Torch is popular among data scientists because it provides a wide range of algorithms and a large user community.

Torch provides two main interfaces: an imperative programming style and a functional style. In the functional style, you first define the mathematical operations (or *nodes*) that make up the graph, and then you "wire" them together. Nodes can be connected in any order, and the resulting graph can be run in any order. This makes it easy to try out new ideas and see what works best.

Torch also has a wide variety of built-in nodes for common tasks, such as matrix operations, convolutions, and so on. You can also write your nodes in C/C++ or Lua.

One of the advantages of Torch is that it is very fast. For example, the popular machine learning algorithm, support-vector machines can be trained much more quickly in Torch than in other languages. This is because Torch uses a library called CUDA, which provides a way to use your computer's graphics processing unit (GPU) on your computer to perform some calculations. This can speed up training by a factor of 10 or more.

Torch is also well-integrated with other scientific computing libraries, such as NumPy and SciPy. This makes it easy to use Torch to perform complex calculations on large data sets. Finally, Torch is open source and free to use.

The Torch package is a library for image augmentation. It provides a range of functions for modifying images, including cropping, rotating, scaling, and flipping. It also includes a range of filters, such as blur and saturation, that can be applied to images.

The Torch package is a useful tool for increasing the diversity of your training data set. By applying a range of image modifications, you can create a more varied set of images that better represent the range of images that your algorithm will encounter in practice. This can help your algorithm to better learn to recognize objects and patterns in images.

The Torch package is also useful for debugging your algorithm. By visually inspecting the output of the torch package, you can quickly identify any problems with your algorithm, which helps you improve the accuracy of your algorithm.

Neural networks are a type of machine learning algorithm that can be used to model complex patterns in data. This means that they can be used to learn how to do things like recognizing animals' faces or understanding speech. The Torch is a library that provides tools for working with neural networks. In this example, you will learn how to create a neural network using R, a programming language.

The codes used in this example are taken from the address "`https://anderfernandez.com/en/blog/how-to-create-neural-networks-with-torch-in-/".r/"`. Once the package is installed, you can create a new matrix data or change the dimensions of existing data.

```
# Install torch package
>install.packages("torch")
# Load the torch library
>library(torch)
# Create a 4x5 random array between 1 and 0.
>torch_rand(4,5)
>torch_tensor
 0.1893  0.6674  0.5244  0.8086  0.4512
 0.1095  0.8961  0.2942  0.9146  0.0603
 0.2876  0.0084  0.3683  0.4768  0.3324
 0.2110  0.5731  0.2986  0.7237  0.2847
[ CPUFloatType{4,5} ]
 # We can convert any matrix to tensor
>x_mat= matrix(c(3,6,1,4,7,9,6,1,4), nrow=3, byrow=TRUE)
>tensor_1=torch_tensor(ex_mat)
>tensor_1
torch_tensor
 3  6  1
 4  7  9
 6  1  4
[ CPUFloatType{3,3} ]
# Conver back to an R object
>my_array=as_array(tensor_1)
```

```
>my_array
     [,1] [,2] [,3]
[1,]    3    6    1
[2,]    4    7    9
```

In our model, we use this model that created by Jauregui 2022, to generate different dimension of wine data set.

```
#Layer 1
>model = model = nn_sequential
nn_linear(13, 20),
nn_relu(),
 # Layer 2
nn_linear(20, 35),
nn_relu(),
 # Layer 3
nn_linear(35,10),
nn_softmax(2)
)
```

The wine data set comes from `https://www.kaggle.com/datasets/nareshbhat/wine-quality-binary-classification`.

```
# Read CSV file
>wine = read.csv('https://archive.ics.uci.edu/ml/machine-learning-
databases/wine/wine.data')
>train_split = 0.75
>sample_indices =sample(nrow(wine) * train_split)
 # 2. Convert our input data to matrices and labels to vectors.
>x_train = as.matrix(wine[sample_indices, -1])
>y_train = as.numeric(wine[sample_indices, 1])
>x_test = as.matrix(wine[-sample_indices, -1])
>y_test = as.numeric(wine[-sample_indices, 1])
# 3. Convert our input data and labels into tensors.
>x_train = torch_tensor(x_train, dtype = torch_float())
>y_train = torch_tensor(y_train, dtype = torch_long())
>x_test = torch_tensor(x_test, dtype = torch_float())
```

```
>y_test = torch_tensor(y_test, dtype = torch_long())
>pred_temp = model(x_train)
>cat(" Dimensions Prediction: ", pred_temp$shape," - Object type
Prediction: ", as.character(pred_temp$dtype), "\n","Dimensions Label: ",
y_train$shape," - Object type Label: ", as.character(y_train$dtype))
>Dimensions Prediction:  132 10  - Object type Prediction:  Float
>Dimensions Label:  132  - Object type Label:  Long>
```

Also, you can resize an image using the "torchvision" packages.

```
# Load torchvision and magick library
>library(torchvision)
>library(magick)
```

The image that used in this example comes from `https://www.kaggle.com/datasets/vishalsubbiah/pokemon-images-and-types`.

```
# Read CSV file
>url_imagen = "C:/Users/Esma/Desktop/sadullah kitap R kodu/archeops.png"
>imagen = image_read(url_imagen)
>plot(imagen)
>title(main = "Original image")
```

The output is shown in Figure 4-9.

Figure 4-9. *Original image*

```
# Draw a chart with 2 rows and 4 columns
>par (mfrow=c(2,2))
>img_width = image_info(imagen)$width
>img_height = image_info(imagen)$height
>imagen_crop = transform_crop(imagen,0,0, img_height/3, img_width/3)
>plot(imagen_crop)
>title(main = "Croped image")
>imagen_crop_center = transform_center_crop(imagen, c(img_height/2,
img_width/2))
>plot(imagen_crop_center)
>title(main = "Croped center image")
>imagen_resize = transform_resize(imagen, c(img_height/5, img_width/5))
>plot(imagen_resize )
>title(main="Resized image")
>imagen_flip = transform_hflip(imagen)
>plot(imagen_flip)
>title(main="Flipped image")
```

The output is shown in Figure 4-10.

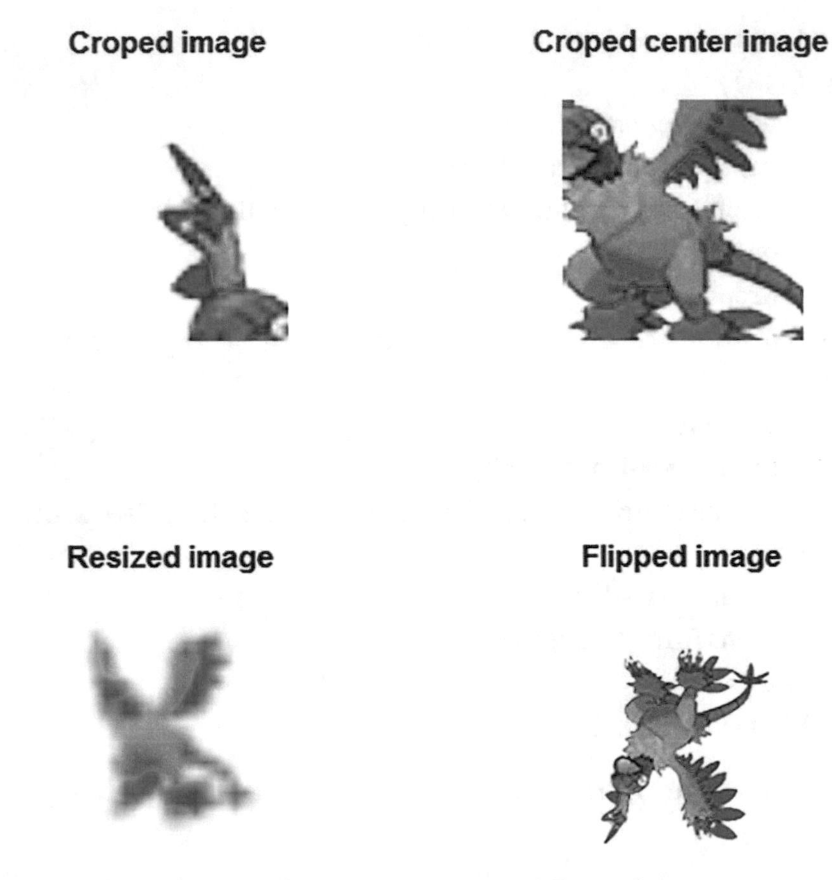

Figure 4-10. *Croped, croped center, resized and flipped image*

Multivariate Imputation Via "mice" Package in R

A synthetic data imputation method is a technique for imputing missing data by creating new, artificial data points that are based on known values. This can be done using a variety of methods, such as creating a regression model to predict missing values or using a clustering algorithm to generate new data points that are similar to existing ones.

The advantage of synthetic data imputation is that it can be used to fill in missing data in a way that is not biased by the existing data. This is because the new data points are generated from scratch, rather than being estimated based on the known data.

The disadvantage of synthetic data imputation is that it can be time-consuming to generate new data points, and there is always the possibility that the artificial data will not be an accurate representation of the real data. In general, synthetic data imputation is a useful technique for handling missing data, but it should be used with caution.

The basis of synthetic data proposed by Rubin (2004) is based on the multiple imputations of missing data. In statistical predictive modeling, missing values affect the reliability of prediction results. Therefore, some common methods are used to solve the missing value problem. Some machine learning algorithms claim to essentially handle missing data. But how good these algorithms are not well known.

Without changing the missing values in the data set, the observed precision values can be modified by making multiple pulls from the non-Bayesian predictive distribution or the posterior predictive distribution. Although real datasets are not used, the multiple synthetic datasets produced can provide accurate statistical inferences. For researchers to draw reliable conclusions from the combined results, analyses using multiple synthetic data must be combined into a single inference. Therefore, the additional variability induced by the imputation procedure needs to be reflected.

The method to be used to assign the missing values greatly affects the success of the predicted model. In most statistical analyzes, the list-based deletion method is used to impute missing values. However, this method is not used much because it causes a loss of information. On the other hand, R software has packages that re successful in imputing missing values. The "mice" "Amelia", "missForest", "Hmisc", and "mi" packages are used to impute missing values in R [4]. There are many ways to impute missing values, each with its advantages and disadvantages. The most important thing is to choose a method that best suits the data set and the type of analysis being performed.

Mice is a package used to impute missing values. This package assumes that missing data is missing randomly. The probability that a value is missing depends only on the observed values. Using these observed values, missing data can be estimated. Mice package uses a method that imputed data on a per-variable basis. This method specifies an assignment pattern per variable.

Usually, the mice package uses joint modeling, sequential modeling, and fully conditional specification methods to impute missing data and write the paragraph above differently. These methods are used to improve the accuracy of the imputation by using information from other variables in the data set. This can be done by using a joint model to impute the missing data and then using a sequential model to improve the accuracy of the imputation. Finally, a fully conditional specification can be used to further improve the accuracy of the imputation.

In the joint modeling method, the joint distribution is specified indirectly; by specifying the distribution of each variable x separately. This method uses a different synthesis model to simultaneously draw synthetic data for all variables in x. In sequential modeling, each variable is factored into a set of univariate models. Each set of models here is conditioned on the variables placed earlier. For example, consider the array $\boldsymbol{x}_{(<j)} = (x_1, \ldots, x_{j-1})$. If this array specifies the variables occurring before x_j, then the $m-th$ synthetic dataset is calculated as follows [5].

$$x_{syn}^{(m)} \sim P(\boldsymbol{x}|\boldsymbol{z}) = \prod_{j=1}^{p} P\left(x_j | \boldsymbol{x}_{(<j)}, \boldsymbol{z}\right)$$

In the fully conditional specification method, each model is separate and can be applied to any other variable in the data set. This is done by conditioning the other variable. For example, let $\boldsymbol{x}_{(-j)} = (x_1, \ldots, x_{j-1}, x_{j+1}, \ldots, x_p)$ specifically, refer to variables in $\boldsymbol{x}$ other than x_j. In this case, the $m-th$ synthetic dataset for each x_j value is drawn using the following formula [5].

$$x_{j,syn}^{(m)} \sim P\left(x_j | \boldsymbol{x}_{(-j)}, \boldsymbol{z}\right) = \prod_{j=1}^{p} P\left(x_j | \boldsymbol{x}_{(<j)}, \boldsymbol{z}\right)$$

In the above equation, it can take $j = 1, 2, \ldots, p$ values. The flexibility that joint modeling, sequential modeling, and fully conditional specification [6] methods provide in determining the joint distribution $P(x, z)$ can sometimes be different. If all variables have a normal distribution and a linear relationship, then the predictive distributions expressed by the three methods are also equivalent.

Now let's make a sample application using the "mice" package.

```
# Load the "mice", "lattice", "dplyr", and "VIM" packages.
>library(mice)
>library(lattice)
>library(dplyr)
>library(VIM)
>set.seed(271)
# Choose the sample size of 200.
>n<-200
# Simulate a random data frame with missing values
> data <- data.frame(sex =  sample (c("F","M")),  age = rnorm(n,15:80), bmi
= rnorm(n,18:50), sbp = rnorm(n, 40:180), dbp = rnorm(n, 50:200), insulin =
rnorm(n,1:50), smoke = rep(c(1, 2), 200))

# Print the first six rows of the dataset
```

```
>head(data)
sex       age       bmi      sbp      dbp    insulin smoke
1   F 13.74044 17.99169 41.50539 47.65656 0.04249654    1
2   M 17.56544 19.17274 40.23685 50.19870 1.77210095    2
3   F 15.48855 19.39617 42.67962 53.53512 2.29891097    1
4   M 18.90000 19.17086 42.46920 52.19573 4.21447491    2
5   F 17.97135 22.85069 44.79011 53.86757 7.50545309    1
6   M 17.06892 22.68086 44.57467 54.38725 7.60687887    2

# Manually add some missing values
> missing.data <- data %>%mutate(age = "is.na<-"(age, age <25 | age >75),
bmi = "is.na<-"(bmi, bmi >44 | bmi <18), sbp = "is.na<-"(sbp, sbp >45 | sbp
<20), dbp = "is.na<-"(dbp, dbp >180 | dbp <65))
> head(missing.data)
  sex age      bmi      sbp dbp    insulin smoke
1   F  NA       NA 41.50539  NA 0.04249654     1
2   M  NA 19.17274 40.23685  NA 1.77210095     2
3   F  NA 19.39617 42.67962  NA 2.29891097     1
4   M  NA 19.17086 42.46920  NA 4.21447491     2
5   F  NA 22.85069 44.79011  NA 7.50545309     1
6   M  NA 22.68086 44.57467  NA 7.60687887     2

# Find the order of missing values in the dataset.
> md.pattern(missing.data)
sex insulin smoke age bmi dbp sbp
200   1       1     1   1   1   1   0   1
10    1       1     1   1   1   0   1   1
32    1       1     1   1   1   0   0   2
44    1       1     1   1   0   1   0   2
16    1       1     1   1   0   0   0   3
22    1       1     1   0   1   1   0   2
14    1       1     1   0   1   0   1   2
24    1       1     1   0   1   0   0   3
32    1       1     1   0   0   1   0   3
2     1       1     1   0   0   0   1   3
4     1       1     1   0   0   0   0   4
      0       0     0  98  98 102 374 672
```

The output is shown in Figure 4-11.

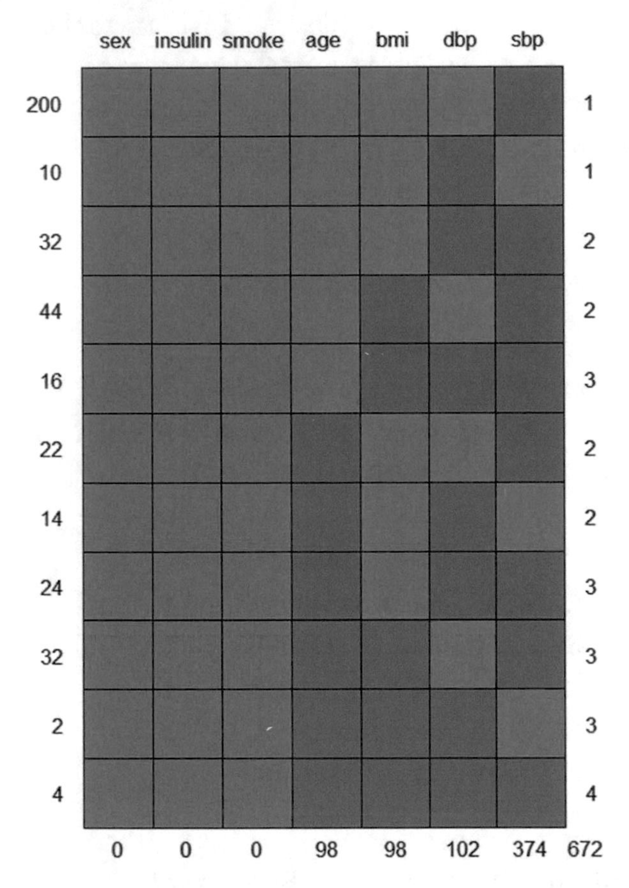

***Figure 4-11.** Missing data pattern*

Blue shows observed values, and red shows missing values. There are 672 missing values in the analyzed data set. Of the missing values, 374 "sbp", 102 "dbp", and 98 "bmi", and "age" belong to the variables. The missing data model is important because it shows how many missing values are in the dataset and how they are distributed.

```
# How many patterns are there where the "bmi" variable is missing.
>mpattern <- md.pattern(missing.data)
> sum(mpattern[, "bmi"] == 0)
[1] 5
There are 5 patterns and 98 cases for the "bmi" variable.
```

Draw the aggr plot graph. The output is shown in Figure 4-12.

```
> aggr_plot <- aggr(missing.data, col=c('red','yellow'), numbers=TRUE,
sortVars=TRUE, labels=names(data), cex.axis=.7, gap=3, ylab=c("Histogram of
missing data","Pattern"))
```

```
Variables sorted by number of missings:
 Variable Count
      sbp 0.935
      dbp 0.255
      age 0.245
      bmi 0.245
      sex 0.000
  insulin 0.000
    smoke 0.000
```

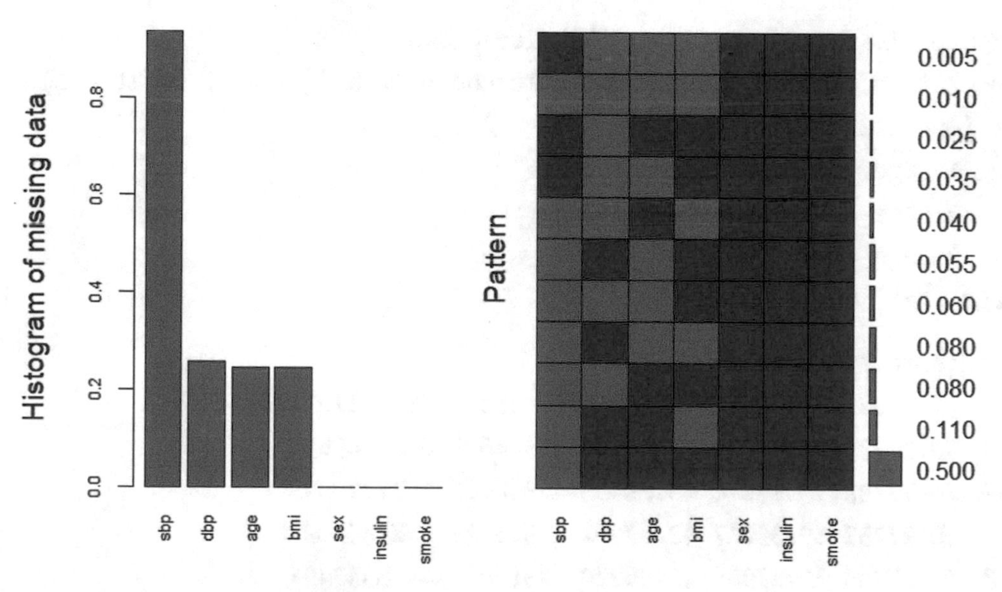

***Figure 4-12.** Histogram of missing data and pattern*

Draw a box plot graph. The output is shown in Figure 4-13.

```
> marginplot(missing.data[c(6,3)])
```

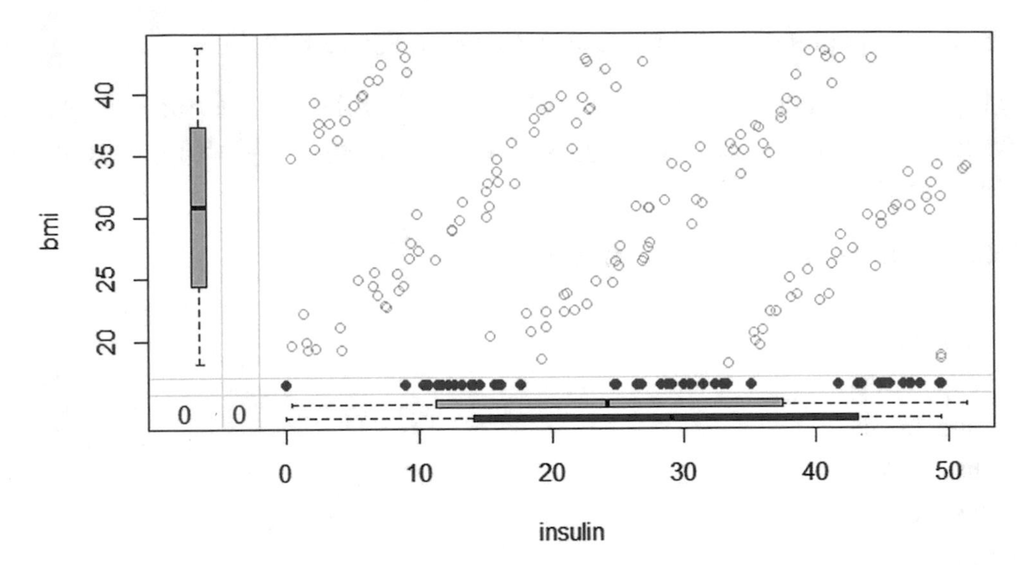

Figure 4-13. *Special box plot of "bmi" and "insulin" variables*

```
# Impute missing values using the mice package
> imputation <- mice(missing.data, method = "mean", m = 4, maxit = 1)
iter imp variable
  1   1  age  bmi  sbp  dbp
  1   2  age  bmi  sbp  dbp
  1   3  age  bmi  sbp  dbp
4  age  bmi  sbp  dbp

> head(complete(imputation))
sex      age      bmi      sbp      dbp    insulin smoke
1   F 50.27733 30.91625 41.50539 113.6025 0.04249654     1
2   M 50.27733 19.17274 40.23685 113.6025 1.77210095     2
3   F 50.27733 19.39617 42.67962 113.6025 2.29891097     1
4   M 50.27733 19.17086 42.46920 113.6025 4.21447491     2
5   F 50.27733 22.85069 44.79011 113.6025 7.50545309     1
6   M 50.27733 22.68086 44.57467 113.6025 7.60687887     2

# Pooling the results and fitting a linear model
>ModelFit <- with(imputation, lm(insulin ~ bmi+age+sbp))
# Combine the results of the 4 models produced
> pool(ModelFit)
Class: mipo    m = 4
```

```
         term m    estimate         ubar b            t
1 (Intercept) 4 26.77349366 7.008515e+03 0 7.008515e+03
2         bmi 4  0.07859746 1.301218e-02 0 1.301218e-02
3         age 4 -0.11191024 3.290485e-03 0 3.290485e-03
4         sbp 4  0.04649987 3.812104e+00 0 3.812104e+00
  dfcom       df riv lambda         fmi
1   396 393.9751   0      0 0.005038099
2   396 393.9751   0      0 0.005038099
3   396 393.9751   0      0 0.005038099
4   396 393.9751   0      0 0.005038099

>summary(pool(ModelFit))

  term      estimate   std.error   statistic       df    p.value
1 (Intercept) 26.77349366 83.71687609  0.31980999 393.9751 0.74928190
2         bmi  0.07859746  0.11407094  0.68902267 393.9751 0.49121457
3         age -0.11191024  0.05736275 -1.95092198 393.9751 0.05177486
4         sbp  0.04649987  1.95246109  0.02381603 393.9751 0.98101141
> densityplot(missing.data$bmi)
```

The output is shown in Figure 4-14.

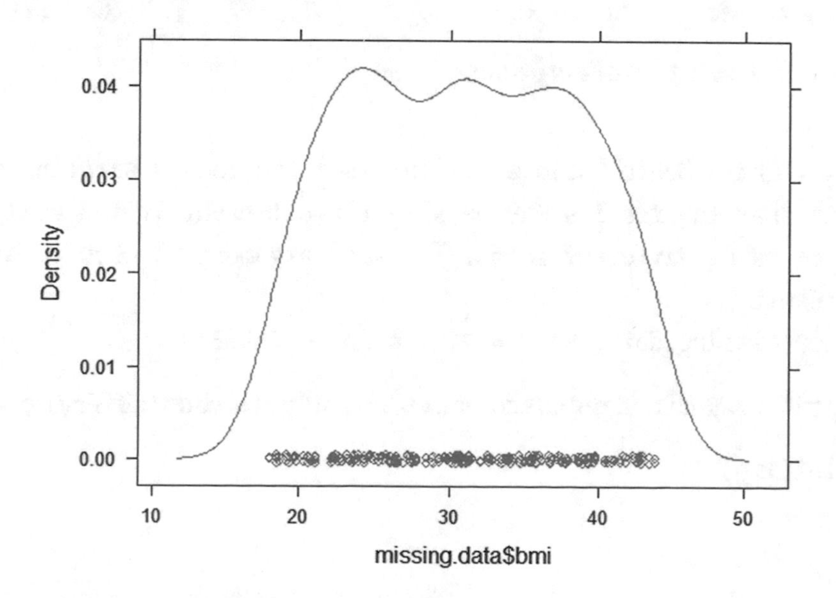

***Figure 4-14.** Density distribution of "bmi" variable*

The "mice" package contains the Markov Chain Monte Carlo algorithm. Images are created using this algorithm.

Plot the imputed density graph. The output is shown in Figure 4-15.

```
>densityplot(imputation)
```

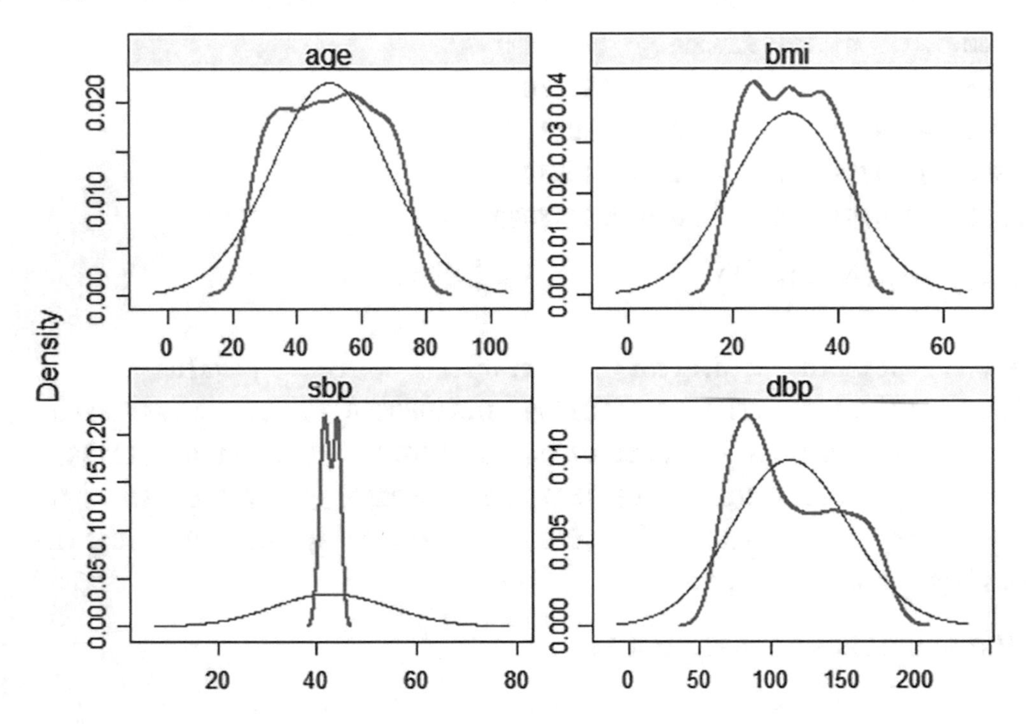

***Figure 4-15.** Density plot of variables*

```
# The Markov Chain Monte Carlo algorithm uses the random sampling method,
which means that the results may be slightly different if the assignments
are repeated using different seeds. The seed argument is used to achieve
the same result.
>imp <- mice(missing.data, seed = 271, print = FALSE)
```

Density plot original and imputed dataset. The output is shown in Figure 4-16.

```
>densityplot(imp)
```

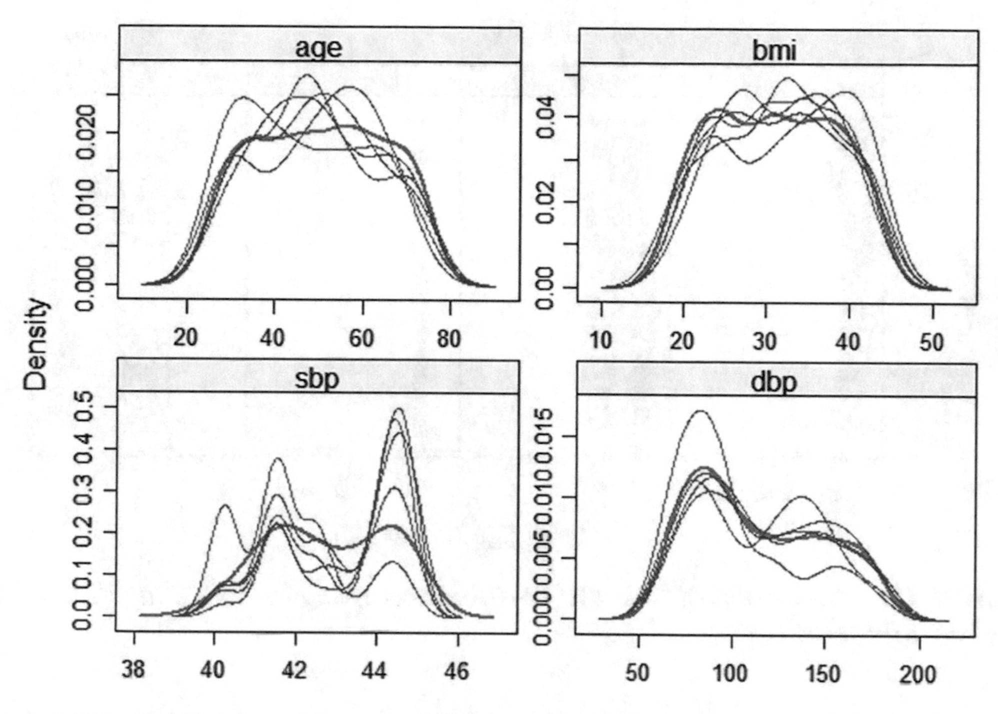

***Figure 4-16.** Original and imputed dataset density plot*

The density of the data imputed in the figure is shown in magenta. The intensity of the observed data is shown in blue. When the figure is examined, it is seen that the distributions of the observed and imputed data are quite similar.

Find the distribution of insulin variable according to other variables. The output is shown in Figure 4-17.

```
> stripplot(imp, insulin ~ bmi+age+sbp+dbp, pch = 3, cex = 0.5)
```

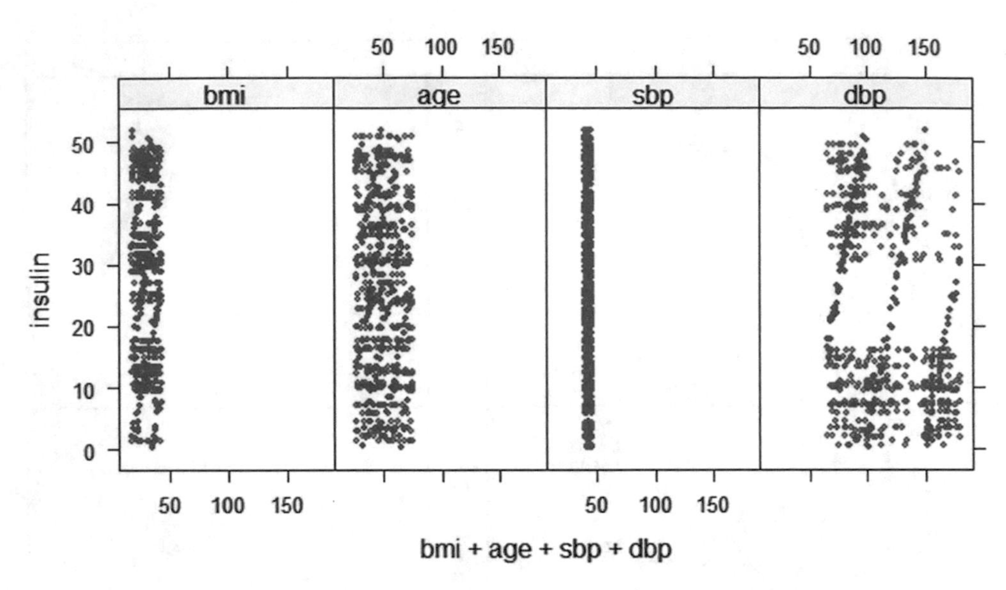

Figure 4-17. *Scatter plot of "insulin" variable according to "bmi", "age", "sbp" and "dbp" variable*

```
# Filter the dataset.
> Orig.Df <- missing.data %>% dplyr::select(age, bmi, sbp, dbp)
>imp1 <- mice(Orig.Df, seed = 271, print = FALSE)
```

Plotting scatterplots of observed and imputed data. The output is shown in Figure 4-18.

```
> stripplot(imp1)
```

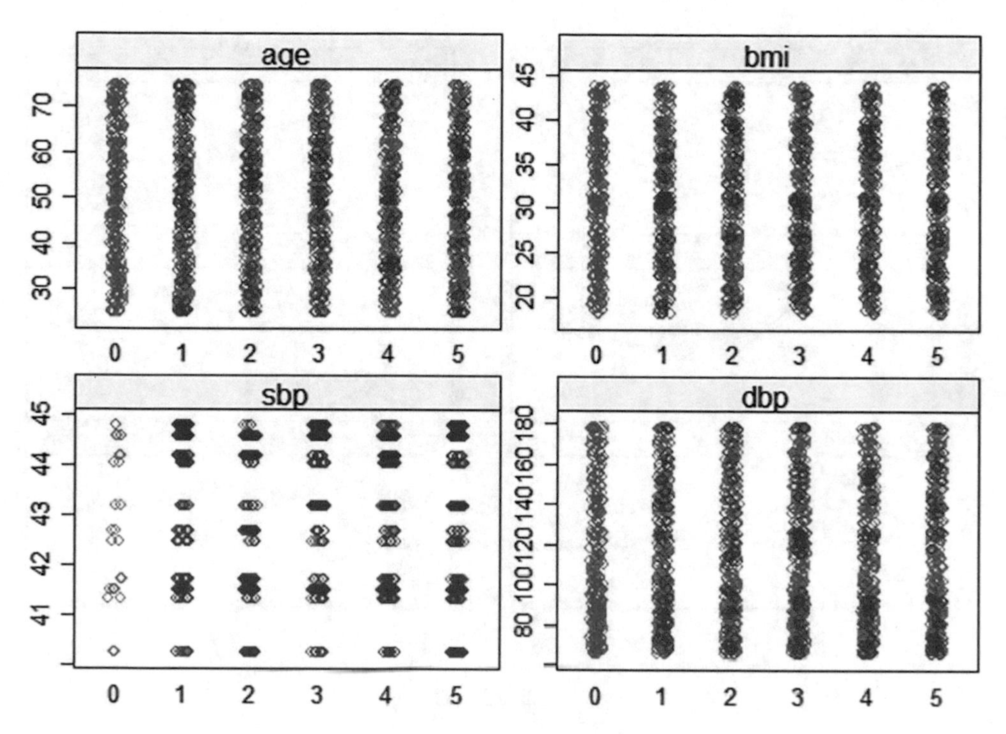

Figure 4-18. *Scatter plot of observed and imputed data*

```
# Check the convergence of the algorithm used
>imp2 <- mice(missing.data)
```

Draw the trace lines of the variables. The output is shown in Figure 4-19.

```
>plot(imp2, c("bmi", "sbp", "dbp"))
```

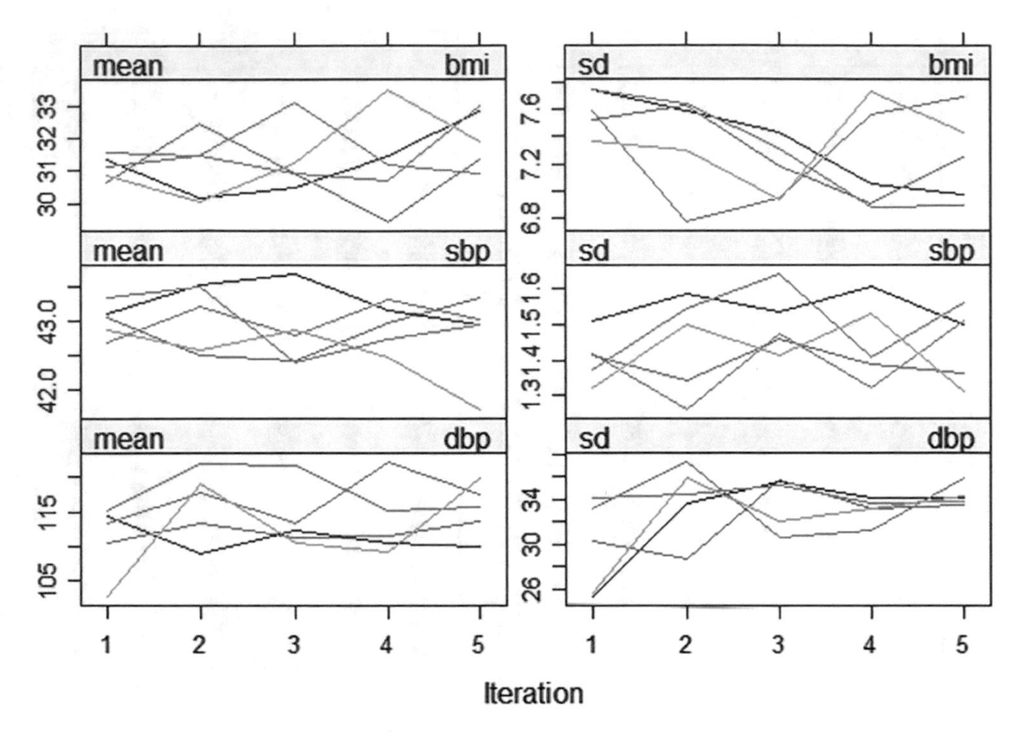

Figure 4-19. *Trace lines of variables*

Generating Synthetic Data with the "conjurer" Package in R

The "conjurer" package in R provides a set of tools for performing mathematical operations on arrays of data. It includes functions for computing the conjugate, determinant, and inverse of matrices, as well as for solving systems of linear equations. The package also includes a variety of other mathematical functions, such as for calculating the sum and product of arrays of numbers.

The "conjurer" package in R is designed for creating synthetic data. It provides several functions for creating data sets with specific distributions, as well as for simulating random events. In this section we will demonstrate how to use the "conjurer" package to generate synthetic data.

The codes used in this example are taken from `https://www.r-bloggers.com/2020/01/generate-synthetic-data-using-r/`. The analysis was done in five steps:

```
# Install the "conjurer" package
> install.packages("conjurer")
```

```
# Load the "conjurer" package
> library(conjurer)
```

Creat a Customer

The customer ID is a number assigned to each customer to help distinguish them from each other. The customer ID is based on how many customers there are and will range from 1 to however many customers there are. The function "buildCust" is used to create the customer ID and it takes a "numOfCust" argument to specify how many customer IDs to create.

```
# Create 1000 customers.
> Customers <- buildCust(numOfCust =  1000)
# View the first six customer IDs.
> head(Customers)
[1] "cust0001" "cust0002" "cust0003" "cust0004" "cust0005" "cust0006"
# Build people names generate 6 person names with a minimum of 4 characters
and a maximum of 9 characters.
>peopleNames <- buildNames(numOfNames =6, minLength =4, maxLength = 9)
>peopleNames
[1] "rayle"     "alien"     "jert"      "manande"   "sharistel" "tarristal"

# Build customer names
> CustomersNames <- as.data.frame(buildNames(numOfNames = 1000, minLength =
8, maxLength = 10))
>head(CustomersNames)
buildNames(numOfNames = 1000, minLength = 8, maxLength = 10)
1                                                  delarleema
2                                                    dalennie
3                                                   colaudiam
4                                                  kinannatre
5                                                   jonicelle
6                                                    jeneliam

# Assign customer name to customer ID
> Customer2Name <- cbind(Customers, CustomersNames)
>head(Customer2Name)
```

```
  Customers buildNames(numOfNames = 1000, minLength = 8, maxLength = 10)
1  cust0001                                                    delarleema
2  cust0002                                                      dalennie
3  cust0003                                                     colaudiam
4  cust0004                                                    kinannatre
5  cust0005                                                     jonicelle
6  cust0006                                                      jeneliam

# Build customer age
> CustomerAge <- as.data.frame(round(buildNum(n = 30, st = 20, en = 70,
disp = 0.5, outliers = 1)))
> colnames(CustomerAge) <- c("CustomerAge ")
> head(CustomerAge)
round(buildNum(n = 30, st = 20, en = 70, disp = 0.5, outliers = 1))
1                                                                    20
2                                                                    25
3                                                                    30
4                                                                    35
5                                                                    40
6                                                                    44
# Assign customer age to customer ID
# Create 30 customers.
> customers <- buildCust(numOfCust =  30)
> Customer2Age <- cbind(customers, CustomerAge)
> head(Customer2Age)
customers round(buildNum(n = 30, st = 20, en = 70, disp = 0.5,
outliers = 1))
1     cust01                                                           20
2     cust02                                                           25
3     cust03                                                           30
4     cust04                                                           35
5     cust05                                                           40
6     cust06                                                           44

# Build customer phone number
>part <- list(c("+90","+33","+45"), c("("), c(505,216,321), c(")"), c(8715:9265))
```

```
> prob <- list(c(0.15,0.20,0.60), c(1), c(0.20,0.50,0.20), c(1), c())
> CustomerPhoneNumbers <- as.data.frame(buildPattern(n=1000,parts = part,
probs = prob))
>head(CustomerPhoneNumbers)
  buildPattern(n = 1000, parts = parts, probs = probs)
1                                         +45(216)8983
2                                         +90(216)8797
3                                         +45(216)9237
4                                         +45(216)8861
5                                         +45(321)9012
6                                         +90(216)9124

> colnames(CustomerPhoneNumbers) <- c("CustomerPhone")
> head(CustomerPhoneNumbers)
CustomerPhone
1  +45(216)8983
2  +90(216)8797
3  +45(216)9237
4  +45(216)8861
5  +45(321)9012
6  +90(216)9124
```

Creat a Product

During product creation, each product is assigned a product ID. The product ID is also similar to the customer ID. The product ID for each product must be between sku001 and sku100, and the price range for each product must be specified long with the product ID. This is a way to create a unique ID for a product. The IDs made up of the product's "sku" number (or any other unique identifier) and the price range for that product. This way, the ID will always be the same length. For example, let's find 20 products between $30-80 using the "buildProb" function. The number of IDs of the products to be created here is shown as "numOfProud", the minimum price is shown with "minPrice", and the maximum price is shown with "maxPrice".

```
# Find 20 items priced between $30 and $80.
> products <- buildProd(numOfProd = 20, minPrice = 30, maxPrice = 80)
# Print 20 products with prices between $30 and $80.
```

```
> products
     SKU Price
1  sku01 73.55
2  sku02 59.68
3  sku03 75.22
4  sku04 52.59
5  sku05 32.61
6  sku06 37.14
7  sku07 48.49
8  sku08 57.15
9  sku09 33.25
10 sku10 30.55
11 sku11 47.97
12 sku12 75.24
13 sku13 39.02
14 sku14 63.18
15 sku15 42.14
16 sku16 70.78
17 sku17 65.41
18 sku18 70.11
19 sku19 47.39
20 sku20 47.66
```

Creating Transactions

Transactions are created after a set of customer IDs and products are created. Transactions are created using the "genTrans" function.

```
# Create a transaction
>Trans <- genTrans(cycles = "m", spike = 5, outliers = 1, transactions
= 10000)
# Visualize the transaction.
> Aggregated <- aggregate(Trans$transactionID, by =
list(Trans$dayNum), length)
> plot(Aggregated, type = "l", ann = FALSE)
```

The output is shown in Figure 4-20.

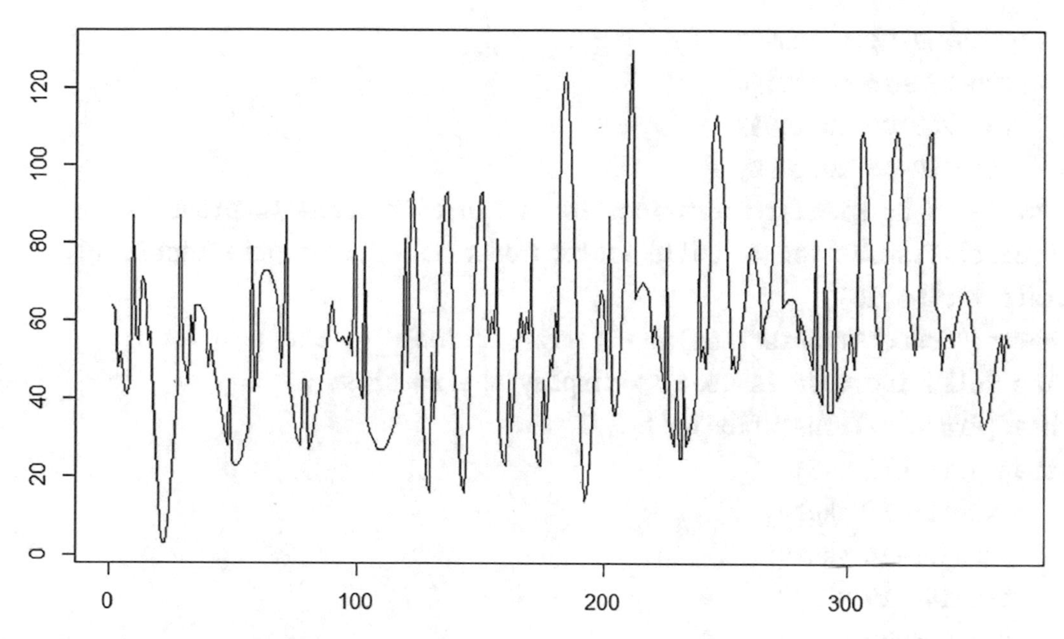

Figure 4-20. *Transaction graph*

Generating Synthetic Data

In the final stage of synthetic data, customers, products, and transactions are brought together. Transactions are created with the "buildPareto" function. This function has "factor1", "factor2" and "Pareto" arguments in its structure. "factor1", and "factor2" are variables to match with each other. The "Pareto" variable is based on the concept of the Pareto principle. Accordingly, the sum of the *x* and *y* values is numerically 100 and is expressed as *c*(*x*,*y*). If Pareto is *c*(80,20), 80% of "factor1" is allocated to 20% of "factor2".

```
# Transactions are allocated to customers using the code below.
> Customer2Transaction <- buildPareto(Customers, Trans$transactionID,
pareto = c(80,20))
# The following code is used to assign readable names to the output.
> names(Customer2Transaction) <- c('transactionID', 'Customer')
# The following code is used to display the output results.
> print(head(Customer2Transaction))
  transactionID Customer
1    txn-222-23 cust0753
2     txn-13-08 cust0523
```

```
3     txn-32-04 cust0900
4    txn-271-36 cust0196
5    txn-208-09 cust0993
6    txn-120-04 cust0909
# Now let's do similar operations to assign operations to products.
> Product2Transaction <- buildPareto(products$SKU,Trans$transactionID,
pareto = c(90,10))
> names(Product2Transaction) <- c('transactionID', 'SKU')
# The following code is used to display the results.
> head(Product2Transaction)
  transactionID   SKU
1    txn-114-26 sku20
2     txn-72-24 sku20
3    txn-184-26 sku20
4    txn-184-39 sku03
5    txn-116-05 sku03
6     txn-44-47 sku20

# Let's assign transactions to products using a similar step to the above
operations.
> Df1 <- merge(x = Customer2Transaction, y = Product2Transaction, by =
"transactionID")
> Now let's create the dataset about transactions, customers and products.
> DfFinal <- merge(x = Df1, y = Trans, by = "transactionID", all.x = TRUE)
# The following code is used to display the dataset.
> head(DfFinal)
  transactionID Customer   SKU dayNum mthNum
1      txn-1-01   cust20 sku20      1      1
2      txn-1-02   cust20 sku03      1      1
3      txn-1-03   cust15 sku20      1      1
4      txn-1-04   cust01 sku20      1      1
5      txn-1-05   cust03 sku20      1      1
6      txn-1-06   cust20 sku20      1      1
```

As a result, a data set of transactions, customers, and products is obtained by recording the transactions of a business over 365 days. This data set contains information on the customers who made the transactions, the products that were purchased, and on which the date the transaction occurred. "transactionID" is a different identifier of the transaction. "SKU" indicates the product purchased in the transaction. "dayNum" indicates how many days in a year. "mthNum" indicates which month it is. If "mthNum" is 1, it represents January, and if it is 12, it represents December.

Generating Synthetic Data with "Synthpop" Package In R

Data synthesis techniques such as synthpop and GANS often use purely conditional synthesis. These techniques use all available information from the model for each variable to find all meaningful relationships between variables. These developed models are quite different from the old methods that used small variable sets selected by hand. If two different subsets of data have very different distribution patterns between immediate variables, it may be difficult for some models to accurately predict both datasets. In this case, positions in the data can be determined using NIST's k-marginal metric.

The "synthpop" package is a tool for synthesizing realistic population data. It can be used to generate synthetic data sets that are representative of a real population or to generate data sets with specific characteristics that are not present in the real population. The package can be used to generate data set with a wide variety of characteristics, including:

- Demographic characteristics (age, sex, race, etc.)
- Socioeconomic characteristics (income, education, occupation, etc.)
- Behavioral characteristics (consumer behavior, health-related behaviors, etc.)
- Spatial characteristics (location, mobility patterns, etc.)

The package can be used to generate data sets of any size, from a few hundred to millions of individuals. The data sets can be generated for any geographic area, from a single city to an entire country.

The "synthpop" package is a valuable tool for researchers and analysts who need realistic synthetic data set. It can generate data sets for simulations or test data-driven models and algorithms. The package can also be used to generate data sets for use in marketing or social science research.

Now let's make a sample application using the "synthpop" package.

```
# Load the "synthpop","tidyverse","sampling" and "partykit" packages.
> library(synthpop)
> library(tidyverse)
> library(sampling)
> library(partykit)
> cols <- c("magenta", "green")
> options(xtable.floating = FALSE)
> options(xtable.timestamp = "")
>my.seed<-235
```

Datasets in many different fields, such as data science and machine learning, have missing values. Therefore, it is often difficult to effectively handle this data, and there is often no best solution. On the other hand, missing data is also important for the model and the target problem. There is no perfect way to handle missing data. Here, it is explained how datasets with missing values should be handled.

The purpose of synthesizing synthetic data is to produce a dataset that is most similar to the original dataset. Two methods are used here to handle missing data: the processing of missing values before analysis, and the use of imputation methods. Here, the imputation method is generally preferred because it gives better results. Because pre-generating missing values from the analysis can affect how missing values are handled.

```
# Select the number of samples to be used in the analysis.
>n<-200
# Simulate a random data frame with 200 samples
> data <- data.frame(sex =  sample (c("F","M")), age = rnorm(n,18:70),
educ = sample(c("primary","secondary","bachelor","master","doctor")), bmi
= rnorm(n,20:30),chol = rnorm(n,60:400), sbp = rnorm(n, 40:180), dbp =
rnorm(n, 40:180), weight = rnorm(n,45:120), smoke = rep(c("yes","no")),
martial =  sample (c("married","singel")), income=rnorm(n,500:7000))
# Print the first 6 lines of the simulate a rondom data
```

```
>head(data)
    sex      age      educ      bmi     chol      sbp
1 female 18.06006    doctor 19.89060 60.02898 39.10903
2   male 18.45238  bachelor 19.71671 60.03026 41.25491
3 female 21.17174    master 20.02189 62.33284 42.64906
4   male 24.37687   primary 22.96372 63.91154 42.46654
5 female 21.18826 secondary 24.04257 64.85323 44.13783
6   male 24.27277    doctor 26.65987 64.09472 41.55372
       dbp   weight smoke martial   income
1 39.55985 45.91745   yes  singel 499.0535
2 40.51360 46.12215    no married 498.5034
3 42.10348 46.04940   yes  singel 503.3760
4 42.71555 47.86138    no married 502.4676
5 44.16469 49.34013   yes  singel 502.7911
6 45.65828 49.56517    no married 504.0946
```

Use the "misForest" package to impute missing values. The "missForest" package in R data sets is a tool for imputing missing values in datasets. It uses a random forest algorithm to create a model to predict missing values in a data set.

```
# Load the "missForest" package
> library(missForest)
```

Simulate a random data frame with missing values

```
# Generate 5% missing values at random
> Data.mis <- prodNA(data, noNA = 0.05)
# Print the first 6 lines of the missing data
>head(Data.mis)
    sex      age      educ      bmi     chol      sbp
1 female 18.06006    doctor 19.89060       NA 39.10903
2   male 18.45238  bachelor 19.71671 60.03026 41.25491
3 female 21.17174    master 20.02189 62.33284 42.64906
4   male 24.37687   primary 22.96372 63.91154 42.46654
5   <NA> 21.18826 secondary 24.04257 64.85323 44.13783
6   male 24.27277    doctor 26.65987 64.09472 41.55372
       dbp   weight smoke martial   income
1 39.55985 45.91745   yes  singel 499.0535
```

```
2 40.51360 46.12215    no married 498.5034
3 42.10348       NA   yes  singel 503.3760
4 42.71555 47.86138  <NA> married 502.4676
5 44.16469 49.34013   yes  singel 502.7911
6 45.65828       NA    no married 504.0946

# Synthesize data.
>Synthesis <- syn(Data.mis, seed=my.seed)
>Synthesis
Call:
($call) syn(data = Data.mis, seed = my.seed)

Number of synthesised datasets:
($m)  1

First rows of synthesised dataset:
($syn)
     sex      age      educ      bmi      chol       sbp       dbp
1   male 24.30063 bachelor 26.57773 222.36616  59.29010  62.15641
2   male 58.28933  primary 25.90100 102.19059  80.33966  80.84342
3 female 46.36684 bachelor       NA 194.18152 172.98788 167.30525
4   male 28.54694  primary 27.73193 229.53406  72.78681  69.58943
5 female 20.30850   doctor 18.70856  60.18121  48.16518  52.01056
6   male       NA     <NA> 26.03880 220.49634  56.63966  55.12595
     weight smoke martial   income
1  57.01487   yes  singel 660.7345
2  85.11660  <NA>  singel 543.7991
3 108.41621    no married 631.4005
4  66.29327   yes    <NA> 671.2441
5  57.52313    no married 502.5906
6        NA   yes  singel       NA
...

Synthesising methods:
($method)
     sex      age     educ      bmi     chol      sbp      dbp
"sample"   "cart"   "cart"   "cart"   "cart"   "cart"   "cart"
```

```
  weight    smoke  martial   income
  "cart"   "cart"   "cart"   "cart"

Order of synthesis:
($visit.sequence)
    sex      age     educ      bmi     chol      sbp      dbp  weight
      1        2        3        4        5        6        7       8
  smoke martial   income
      9       10       11

Matrix of predictors:
($predictor.matrix)
        sex age educ bmi chol sbp dbp weight smoke martial income
sex       0   0    0   0    0   0   0      0     0       0      0
age       1   0    0   0    0   0   0      0     0       0      0
educ      1   1    0   0    0   0   0      0     0       0      0
bmi       1   1    1   0    0   0   0      0     0       0      0
chol      1   1    1   1    0   0   0      0     0       0      0
sbp       1   1    1   1    1   0   0      0     0       0      0
dbp       1   1    1   1    1   1   0      0     0       0      0
weight    1   1    1   1    1   1   1      0     0       0      0
smoke     1   1    1   1    1   1   1      1     0       0      0
martial   1   1    1   1    1   1   1      1     1       0      0
income    1   1    1   1    1   1   1      1     1       1      0

# Compare the synthesized data using the "compare" function.
>compare(Synthesis, Data.mis, nrow = 3, ncol = 4, cols = cols)$plot
[1] TRUE

# Choose the variables.
> ods1 <- Data.mis[ , c("age", "bmi", "weight", "sbp", "income")]
> syn1 <- syn(ods1, cont.na = list(income=-6))
Synthesis
-----------
 age bmi weight sbp income
```

Compare data distributions using the Histogram Similarity method. The output is shown in Figure 4-21.

```
>compare(syn1$syn, ods1, vars = "bmi")
Comparing percentages observed with synthetic
Selected utility measures:
        pMSE   S_pMSE df
bmi 0.001689 1.081081  5
```

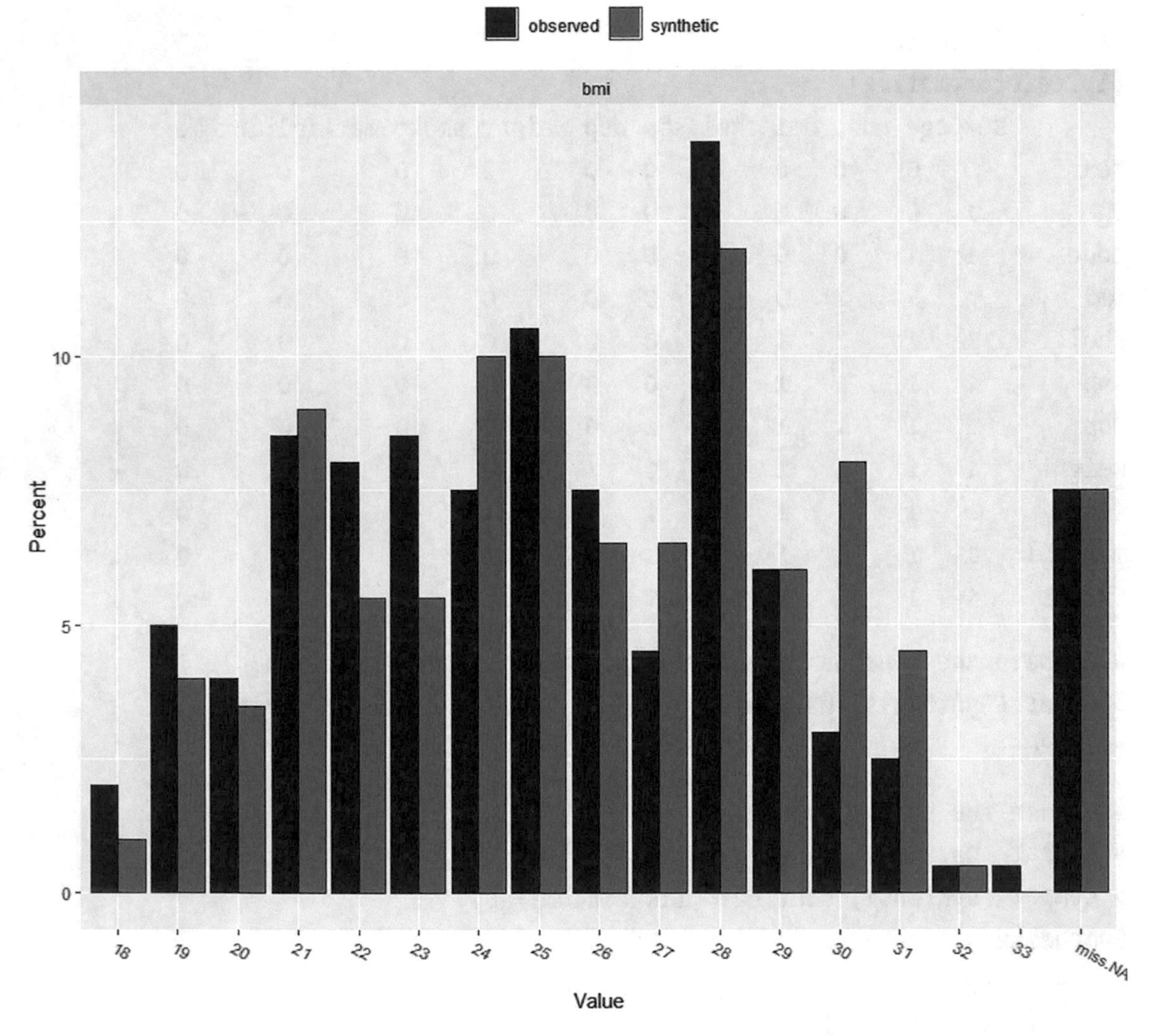

Figure 4-21. *Histogram of observed and synthetic data for the bmi variable*

Considering the above histogram, the observed and synthetic data distributions corresponding to each column largely overlap.

```
# Selecting variables.
> ods2 <- Data.mis[ , c("age", "bmi", "weight", "dbp", "income")]
>syn2 <- syn(ods2, cont.na = list(bmi=-6))
Synthesis
-----------
 age bmi weight dbp income
>sds2 <- syn(ods2, method = "ctree", m = 6)
Synthesis number 1
--------------------
 age bmi weight dbp income
Synthesis number 2
--------------------
 age bmi weight dbp income
Synthesis number 3
--------------------
 age bmi weight dbp income
Synthesis number 4
--------------------
 age bmi weight dbp income
Synthesis number 5
--------------------
 age bmi weight dbp income
Synthesis number 6
--------------------
 age bmi weight dbp income
# Compare data distributions using the Histogram Similarity method.
> compare(sds2, ods2, vars = "weight", msel = 1:3)
Comparing percentages observed with synthetic

Selected utility measures:
          pMSE   S_pMSE df
weight 0.00167 1.068513  5
```

According to Figure 4-22, a large overlap of 3 different synthetic data distributions was produced with the original data corresponding to each column.

```
# Compare the data distributions produced for the "income" variable with
the Histogram Similarity method.
> compare(syn2$syn, ods2, vars = "income", cont.na = list(income = -6),
stat = "counts", table = TRUE, breaks = 10)
```

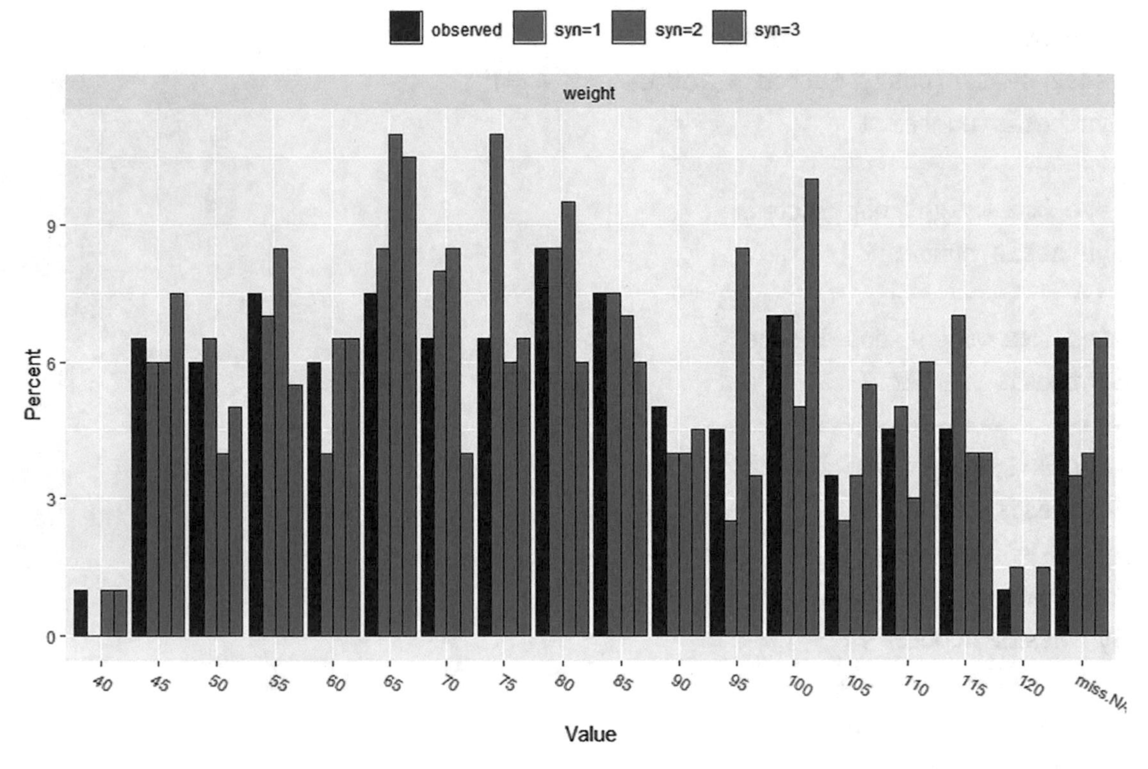

***Figure 4-22.** Histogram of observed and synthetic data for the weight variable*

Comparing counts observed with synthetic. The output is shown in Figure 4-23.

```
$income
          480 500 520 540 560 580 600 620 640 660 680 700 miss.NA
observed    2  19  17  20  18  19  19  20  19  21  15   1      10
synthetic   2  13  15  23  20  11  17  25  24  22  15   1      12

Selected utility measures:
           pMSE   S_pMSE df
income 0.001549 0.991344  5
```

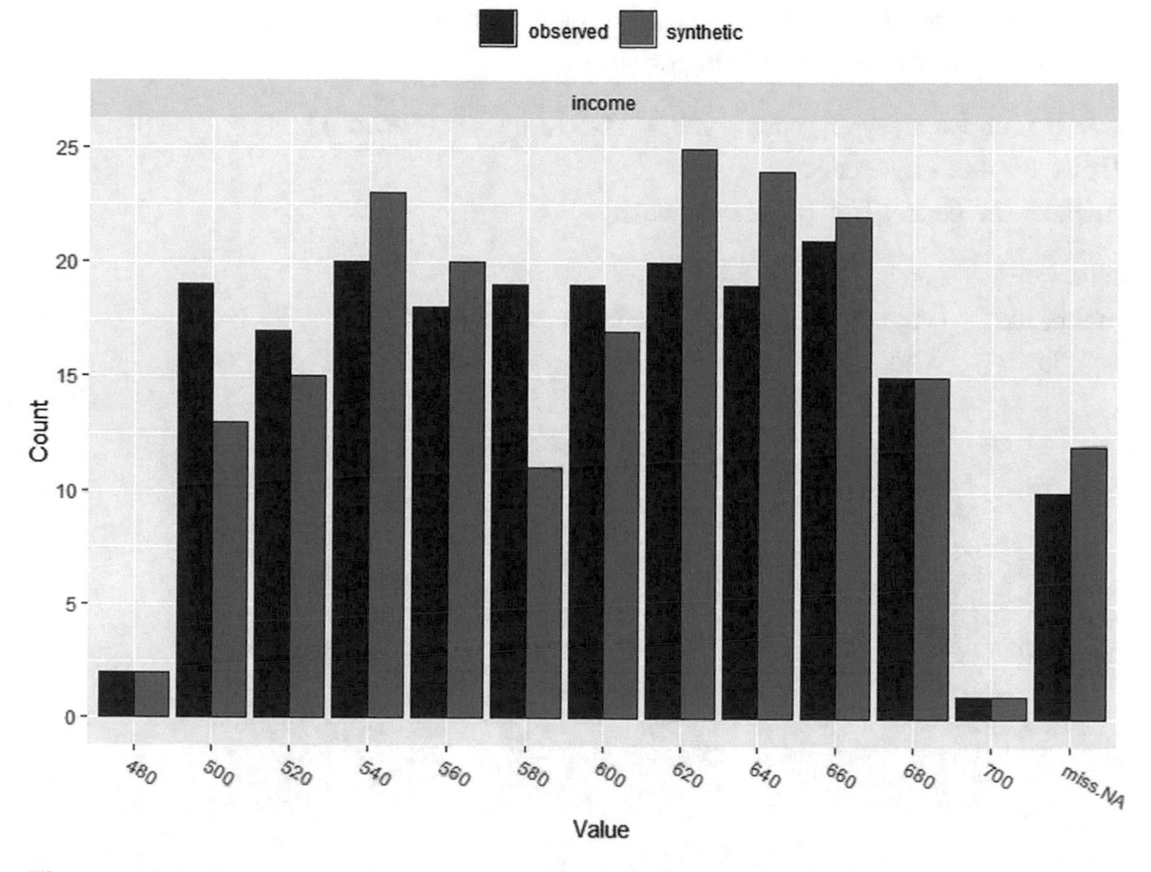

Figure 4-23. *Frequency distribution of "income" variable for observed and synthetic data*

```
# Selecting and synthesize the variables.
> vars3 <- c("sex", "age", "educ", "income", "smoke")
> ods3 <- na.omit(Data.mis[1:500, vars3])
> syn3 <- syn(ods3)
Variable(s): sex, educ, smoke have been changed for synthesis from
character to factor.

Variables sex, smoke are collinear. Variables later in 'visit.sequence'
are derived from sex.

Synthesis
-----------
 sex age educ income smoke
```

Compare the original data with the synthetic data using the Future Importance method. The output is shown in Figure 4-24.

```
> multi.compare(syn3, ods3, var = "sex", by = c("educ"))
Plots of sex  by  educ
Numbers in each plot (observed data):

educ
 bachelor    doctor    master   primary secondary
     30        30        33        30        34
```

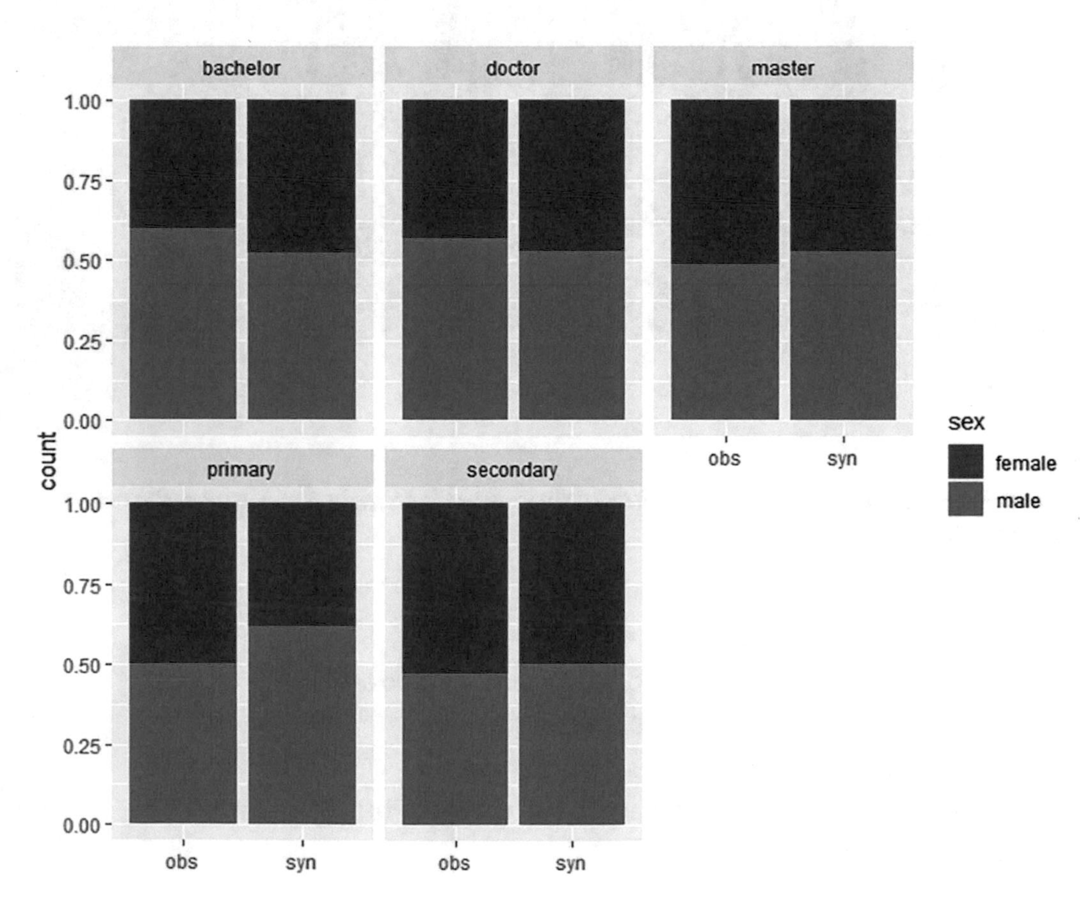

***Figure 4-24.** Comparison of education level by gender*

```
# Compare the original data with the synthetic data using the Future
Importance method.
```

Multiple comparison: The output is shown in Figure 4-25.

```
> multi.compare(syn3, ods3, var = "smoke", by = c("sex","educ"))
Plots of smoke  by  sex educ
Numbers in each plot (observed data):
        educ
sex      bachelor doctor master primary secondary
  female       12     13     17      15        18
  male         18     17     16      15        16
```

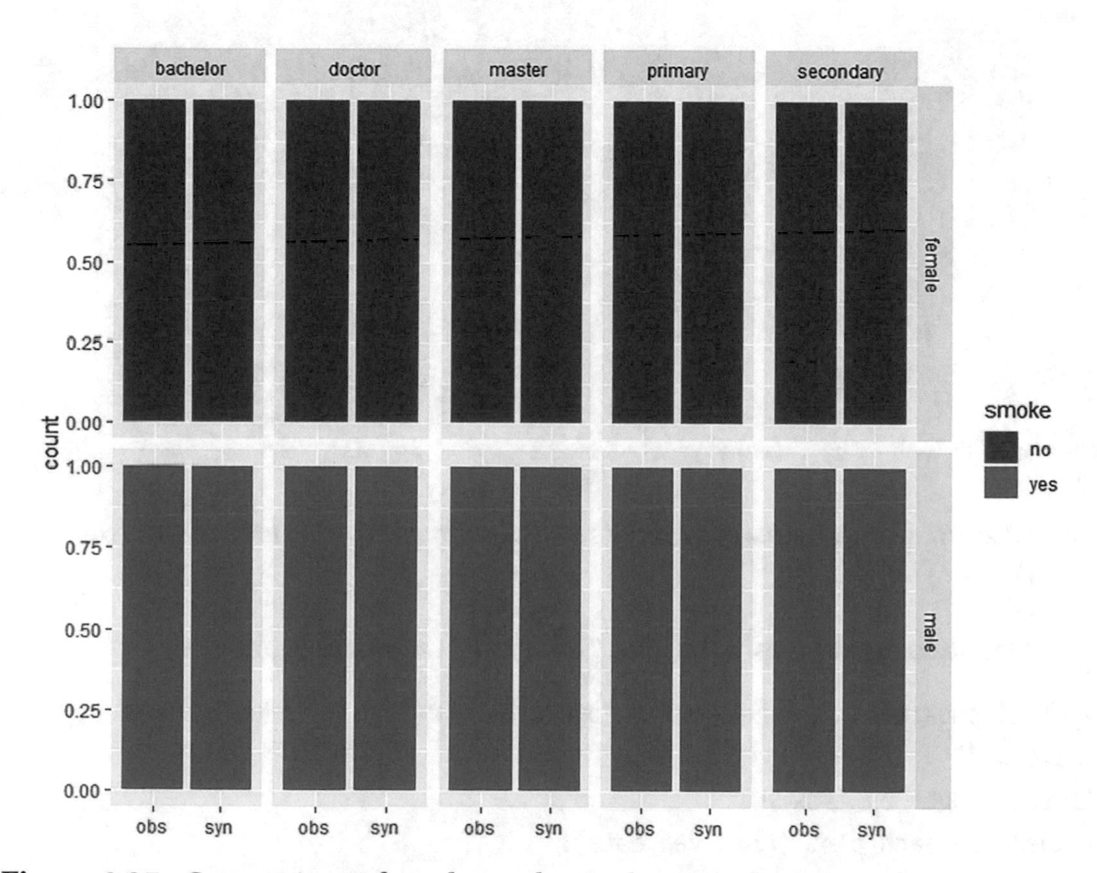

***Figure 4-25.** Comparison of smoke use by gender and education*

Multiple comparison using the Histogram Similarity method. The output is shown in Figure 4-26.

```
> multi.compare(syn3, ods3, var = "age", by = c("sex", "educ"), y.hist =
"density", binwidth = 5)
Plots of age  by  sex educ
Numbers in each plot (observed data):
```

```
         educ
sex       bachelor doctor master primary secondary
  female        12     13     17      15        18
  male          18     17     16      15        16
```

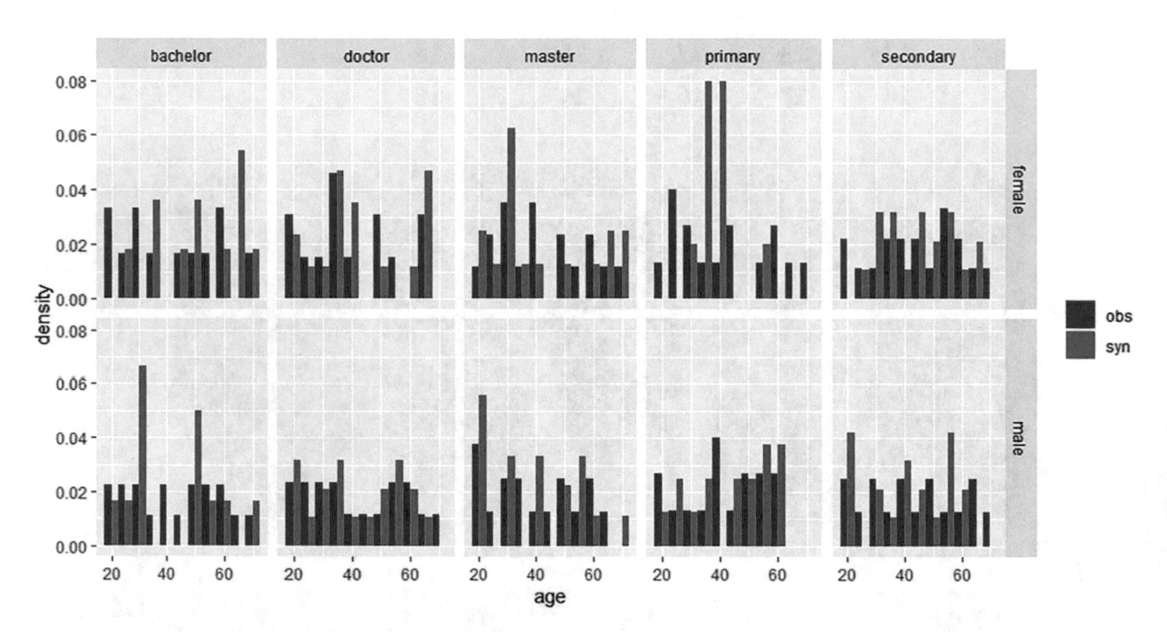

***Figure 4-26.** Comparison of synthetic and observed datasets for sex and educ variables*

Multiple comparison using boxplot. The output is shown in Figure 4-27.

```
>multi.compare(syn3, ods3, var = "age", by = c("sex", "educ"), cont.type =
"boxplot")
Plots of age  by  sex educ
Numbers in each plot (observed data):

         educ
sex       bachelor doctor master primary secondary
  female        12     13     17      15        18
  male          18     17     16      15        16
```

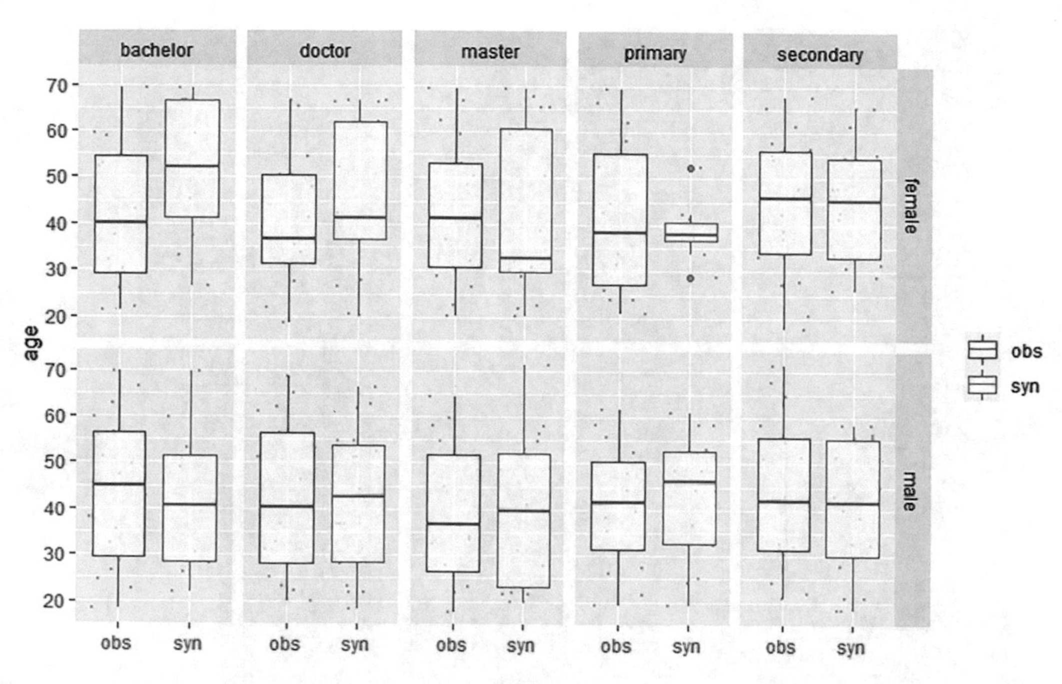

Figure 4-27. *Multiple comparison with boxplot*

As a result, the distributions were similar, and very good results were obtained.

```
>multi.compare(syn3, ods3, var = "income", by = c("smoke"), cont.type =
"boxplot")
```

Plots of income by smoke. The output is shown in Figure 4-28.

```
Numbers in each plot (observed data):
smoke
 no yes
 75  82
```

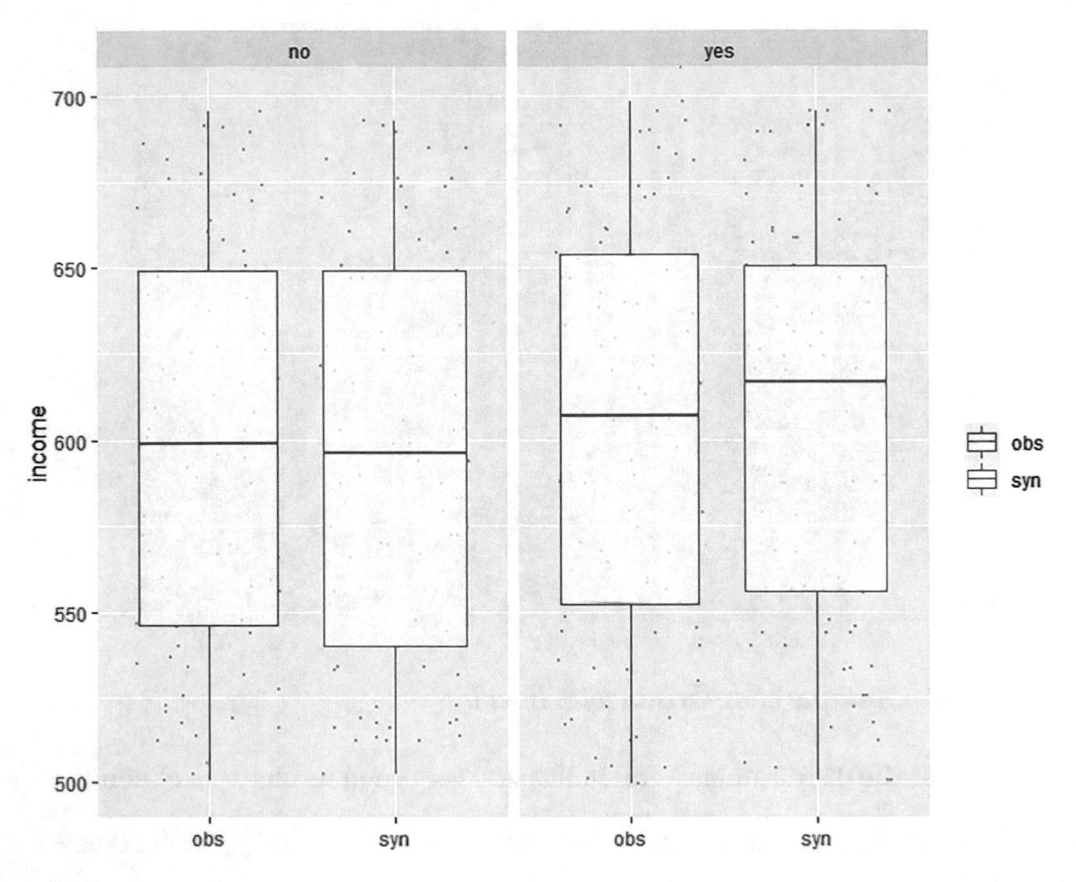

Figure 4-28. *Comparison with boxplot*

Example: Linear model

```
Compare model estimates based on synthesized and observed data
# Select variables
>ods4 <- Data.mis[,c("age","bmi","chol","sbp", "dbp", "weight", "income")]
>ods4$income[ods4$income == -8] <- NA
>syn4 <- syn(ods4, m = 3)
Synthesis number 1
--------------------
 age bmi chol sbp dbp weight income
Synthesis number 2
--------------------
 age bmi chol sbp dbp weight income
Synthesis number 3
```

```
--------------------
 age bmi chol sbp dbp weight income
# Modeling and compare results.
>f1 <- lm.synds(income ~ age + bmi + sbp + dbp, data = syn4)
>f1
Call:
lm.synds(formula = income ~ age + bmi + sbp + dbp, data = syn4)

Average coefficient estimates from 3 syntheses:
 (Intercept)          age          bmi          sbp          dbp
580.36549727   0.05420626  -0.15926752   0.30738785  -0.08043258
> print(f1, msel = 1:3)
Call:
lm.synds(formula = income ~ age + bmi + sbp + dbp, data = syn4)

Coefficient estimates for selected synthetic data set(s):
      (Intercept)        age       bmi         sbp       dbp
syn=1    524.5111  0.6366150 1.2367056 -1.84179575 2.0765711
syn=2    510.2192  0.9555930 0.2355976 -0.96828566 1.3489566
syn=3    578.9077 -0.1032436 -0.3409441  0.02424215 0.2715823

> summary(f1)
Fit to synthetic data set with 3 syntheses. Inference to coefficients
and standard errors that would be obtained from the original data.
Call:
    lm.synds(formula = income ~ age + bmi + sbp + dbp, data = syn4)

Combined estimates:
            xpct(Beta) xpct(se.Beta) xpct(z) Pr(>|xpct(z)|)
(Intercept) 559.680987      39.987390 13.9964         <2e-16 ***
age           0.092128       0.317455  0.2902         0.7717
bmi          -0.033654       1.451448 -0.0232         0.9815
sbp          -0.172531       1.250066 -0.1380         0.8902
dbp           0.528393       1.228929  0.4300         0.6672
---
Signif. codes:  0 '***' 0.001 '**' 0.01 '*' 0.05 '.' 0.1 ' ' 1

# Compare model estimates based on synthesised and observed data.
```

```
>compare(f1, ods4, lcol=cols)
Call used to fit models to the data:
lm.synds(formula = income ~ age + bmi + sbp + dbp, data = syn4)

Differences between results based on synthetic and observed data:
               Synthetic    Observed       Diff Std. coef diff CI overlap
(Intercept) 559.68098675 555.8399827  3.8410041     0.09510782  0.9757374
age           0.09212785   0.3703575 -0.2782296    -0.88560583  0.7740760
bmi          -0.03365357   0.3783328 -0.4119863    -0.27807429  0.9290614
sbp          -0.17253127   2.1445340 -2.3170653    -0.64007112  0.8367136
dbp           0.52839349  -1.9439398  2.4723333     0.68386318  0.8255419

Measures for 3 syntheses and 5 coefficients
Mean confidence interval overlap:  0.868226
Mean absolute std. coef diff:  0.5165445

Mahalanobis distance ratio for lack-of-fit (target 1.0): 2
Lack-of-fit test: 10.0003; p-value 0.0752 for test that synthesis model is
compatible
with a chi-squared test with 5 degrees of freedom.
```

Confidence interval plot: The output is shown in Figure 4-29.

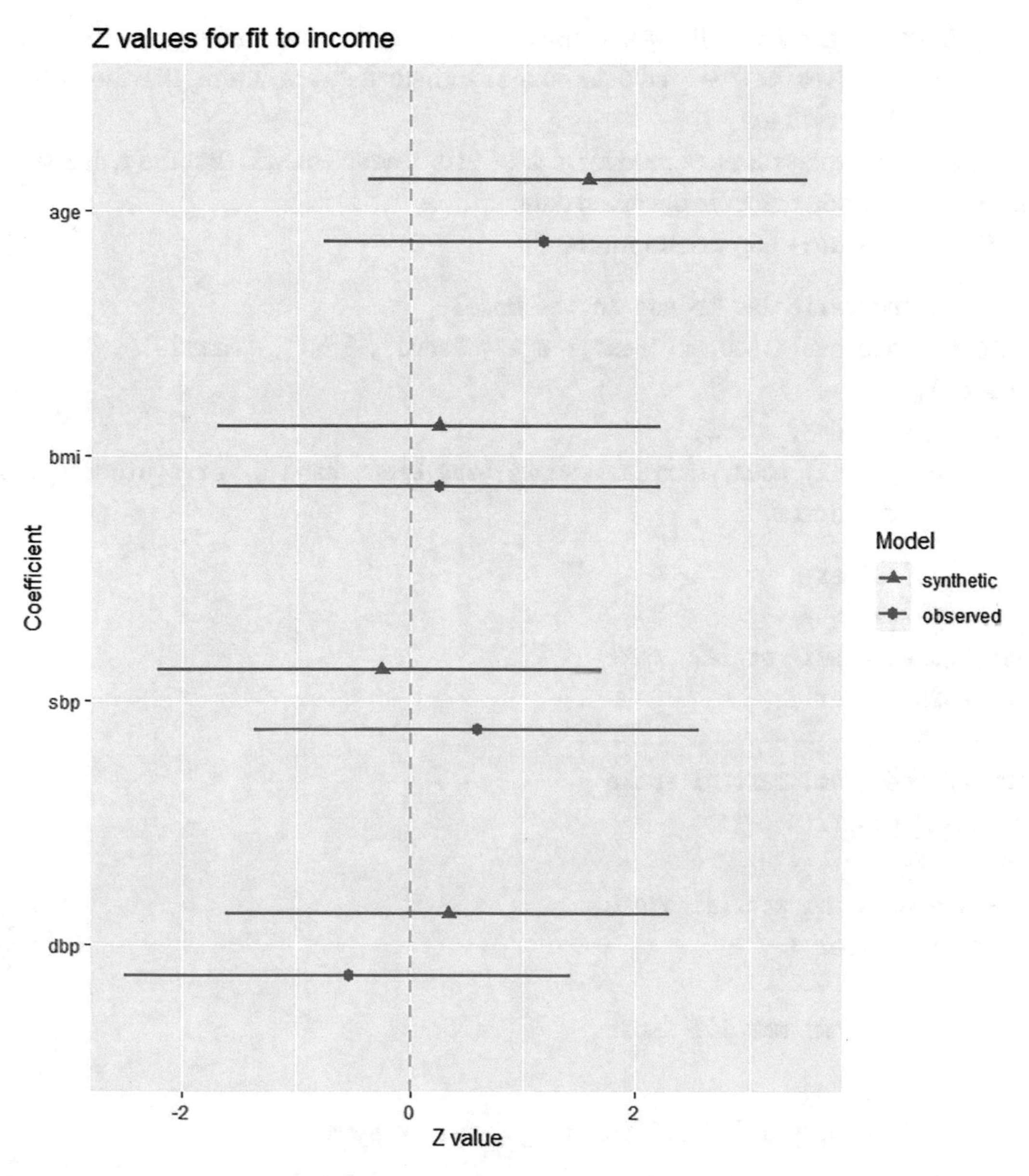

Figure 4-29. *Estimates and 95% confidence intervals for Z statistics from income regression*

The results of the analysis suggest that the model should be refined. Sometimes there may be some contradictory results in the observed data. This situation can be corrected by using different data synthesis methods.

The confidence interval plot in Figure 4-29 shows that the confidence intervals for coefficients estimated using synthetic data are in general, narrower than those estimated using observed data. This is likely because the synthetic data were generated using the model that was fit to the observed data, so the synthetic data are "more" like the data that the model was fit to.

The Mahalanobis distance ratio for lack-of-fit is 2, which means that the synthesis model is not a good fit for the observed data.

Example: Multi-comparison model

```
# Select the variables to use in the model
>ods5 <- Data.mis[1:500, c("sex", "age", "educ", "bmi", "martial",
"smoke")]
>syn5 <- syn(ods5, m = 4)
Variable(s): sex, educ, martial, smoke have been changed for synthesis from
character to factor.

Synthesis number 1
--------------------
 sex age educ bmi martial smoke
Synthesis number 2
--------------------
 sex age educ bmi martial smoke
Synthesis number 3
--------------------
 sex age educ bmi martial smoke
Synthesis number 4
--------------------
 sex age educ bmi martial smoke

# Multi-modelling
>f2 <- multinom.synds(educ ~ sex + age, data = syn5)
>summary(f2)
Fit to synthetic data set with 4 syntheses. Inference to coefficients
and standard errors that would be obtained from the original data.

Call:
multinom.synds(formula = educ ~ sex + age, data = syn5)

Combined estimates:
                        xpct(Beta) xpct(se.Beta) xpct(z) Pr(>|xpct(z)|)
```

```
doctor:(Intercept)       0.4551469     0.8637705  0.5269          0.5982
doctor:sexmale          -0.4650445     0.5351218 -0.8690          0.3848
doctor:age              -0.0039122     0.0188005 -0.2081          0.8352
master:(Intercept)       0.2536342     0.8431040  0.3008          0.7635
master:sexmale          -0.2116635     0.5131161 -0.4125          0.6800
master:age               0.0015296     0.0179901  0.0850          0.9322
primary:(Intercept)      0.0050908     0.8328898  0.0061          0.9951
primary:sexmale         -0.0447069     0.5032999 -0.0888          0.9292
primary:age              0.0073538     0.0175772  0.4184          0.6757
secondary:(Intercept) -0.0449791     0.8402186 -0.0535          0.9573
secondary:sexmale       -0.1789214     0.5080330 -0.3522          0.7247
secondary:age            0.0094189     0.0177082  0.5319          0.5948

>print(f2, msel = 1:3)
Note: To get more details of the fit see vignette on inference.

Call:
multinom.synds(formula = educ ~ sex + age, data = syn5)

Coefficient estimates for selected synthetic data set(s):
      doctor:(Intercept) doctor:sexmale  doctor:age master:(Intercept)
syn=1          0.8784221     -0.4096250 -0.01832298          0.8062422
syn=2          0.9028755     -0.7222305 -0.01085432          0.6584914
syn=3         -0.3549103     -0.6354402  0.01608798         -1.3491940
      master:sexmale   master:age primary:(Intercept) primary:sexmale
syn=1     -0.3531138 -0.005725565          0.35467122     -0.06550066
syn=2     -0.0276484 -0.010354886         -0.03452948      0.07195871
syn=3     -0.1513041  0.031673372         -0.20666485      0.08458827
       primary:age secondary:(Intercept) secondary:sexmale secondary:age
syn=1 -0.002083643             0.3091911       0.008933527  0.0029018830
syn=2  0.010236153             0.6938888      -0.582058180 -0.0008996945
syn=3  0.008444159            -1.3145215      -0.293551212  0.0351327092

#Comparison of model predictions based on generated and observed data.
>compare(f2, ods5, lcol=cols)
# weights:  20 (12 variable)
initial  value 276.823321
iter  10 value 275.777763
```

```
final  value 275.769939
converged

Call used to fit models to the data:
multinom.synds(formula = educ ~ sex + age, data = syn5)

Differences between results based on synthetic and observed data:
                        Synthetic      Observed         Diff
doctor:(Intercept)     0.455146864 -1.420474e-01  0.597194311
doctor:sexmale        -0.465044513 -3.149206e-01 -0.150123952
doctor:age            -0.003912182  6.167936e-03 -0.010080118
master:(Intercept)     0.253634211  1.352540e-03  0.252281671
master:sexmale        -0.211663536  5.720359e-02 -0.268867122
master:age             0.001529574 -3.314822e-05  0.001562722
primary:(Intercept)    0.005090824 -2.955364e-01  0.300627258
primary:sexmale       -0.044706867 -1.769824e-01  0.132275538
primary:age            0.007353846  1.095656e-02 -0.003602717
secondary:(Intercept) -0.044979111  2.573445e-01 -0.302323600
secondary:sexmale     -0.178921424 -2.227352e-01  0.043813788
secondary:age          0.009418886 -8.268434e-04  0.010245730
                      Std. coef diff CI overlap
doctor:(Intercept)        0.75679589  0.8069363
doctor:sexmale           -0.30110724  0.9231855
doctor:age               -0.59778644  0.8475007
master:(Intercept)        0.32548186  0.9169674
master:sexmale           -0.54945873  0.8598294
master:age                0.09392280  0.9760397
primary:(Intercept)       0.38753996  0.9011359
primary:sexmale           0.27375950  0.9301621
primary:age              -0.22001958  0.9438715
secondary:(Intercept)    -0.40100137  0.8977019
secondary:sexmale         0.09131598  0.9767047
secondary:age             0.62787823  0.8398240

Measures for 4 syntheses and 12 coefficients
Mean confidence interval overlap:  0.9016549
Mean absolute std. coef diff:  0.3855056
```

```
Mahalanobis distance ratio for lack-of-fit (target 1.0): 1.14
Lack-of-fit test: 13.64079; p-value 0.3242 for test that synthesis model is
compatible
with a chi-squared test with 12 degrees of freedom.
```

Confidence interval plot: You will see the graph shown in Figure 4-30.

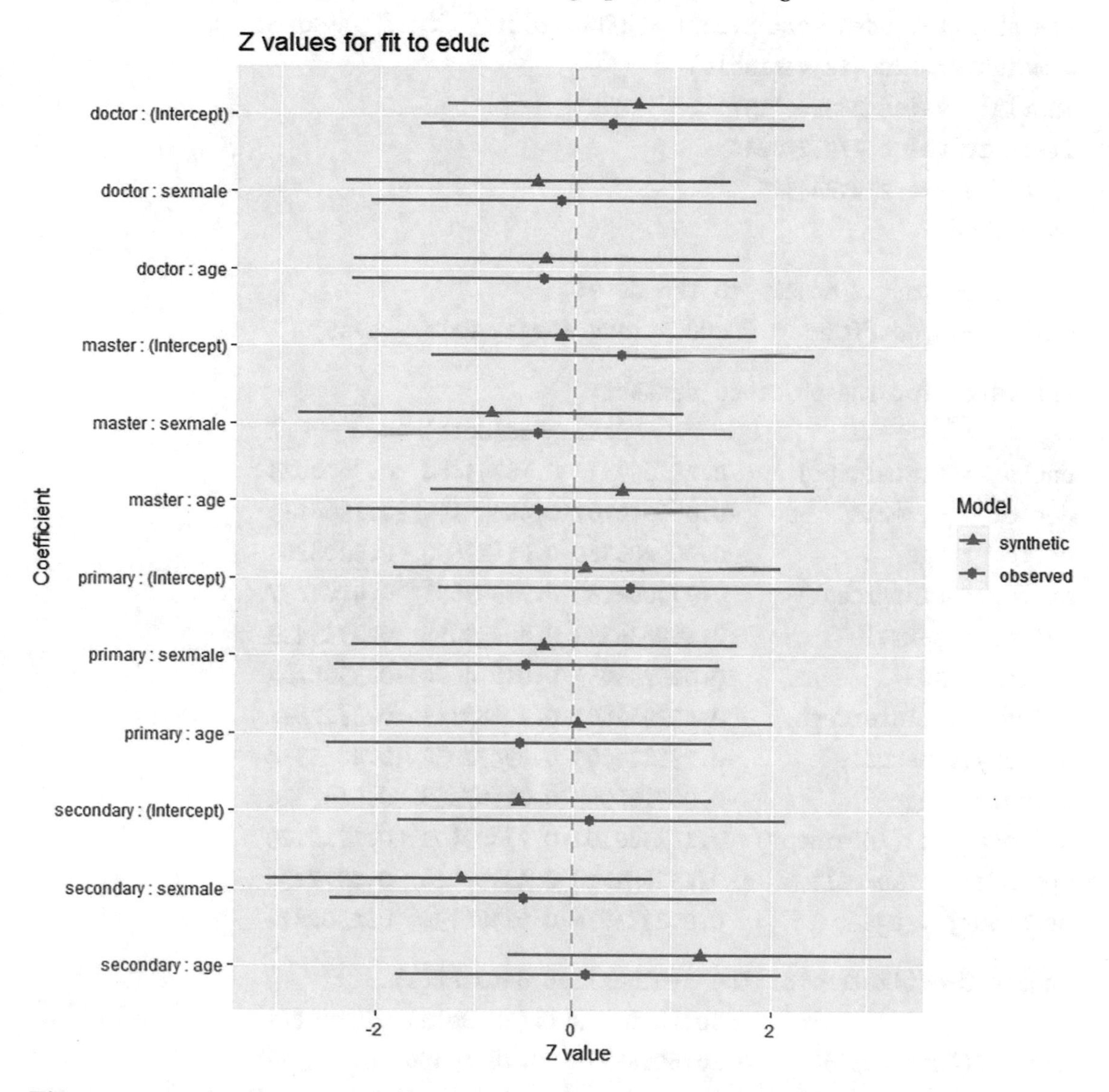

Figure 4-30. *Estimates and 95% confidence intervals for Z statistics from educ regression*

The Mahalanobis distance inequality test shows that the agreement between the synthetic data and the observational data is tested with a chi-square test. The R analysis output above shows that the average of 4 syntheses and 12 coefficients is 0.9016549. This shows that the synthesis models agree with the real data.

```
#Comparing synthetic and observed data
>compare(f2, ods5, print.coef = TRUE, plot = "coef", lcol=cols)
# weights:  20 (12 variable)
initial  value 280.042197
iter  10 value 279.270244
final  value 279.244345
converged

Call used to fit models to the data:
multinom.synds(formula = educ ~ sex + age, data = syn5)

Estimates for the observed dataset:
                                Beta   se(Beta)          Z
doctor : (Intercept)       0.287650834 0.78249484  0.3676073
doctor : sexmale          -0.066070707 0.48433999 -0.1364139
doctor : age              -0.005206560 0.01622890 -0.3208202
master : (Intercept)       0.373062189 0.77993658  0.4783237
master : sexmale          -0.180835533 0.48359055 -0.3739435
master : age              -0.005793053 0.01624404 -0.3566264
primary : (Intercept)      0.452962803 0.79288115  0.5712871
primary : sexmale         -0.238224469 0.49472054 -0.4815334
primary : age             -0.009166959 0.01667258 -0.5498224
secondary : (Intercept)    0.132380101 0.77409402  0.1710129
secondary : sexmale       -0.235881218 0.47392024 -0.4977234
secondary : age            0.002383676 0.01588194  0.1500872

Combined estimates for the synthesised dataset(s):
                        xpct(Beta) xpct(se.Beta)    xpct(z)
doctor:(Intercept)      0.167698852    0.78249484  0.2143130
doctor:sexmale          0.406756817    0.48433999  0.8398167
doctor:age             -0.008294055    0.01622890 -0.5110668
master:(Intercept)      0.925523568    0.77993658  1.1866652
```

```
master:sexmale         -0.103747550    0.48359055 -0.2145359
master:age             -0.018554636    0.01624404 -1.1422429
primary:(Intercept)     0.169180760    0.79288115  0.2133747
primary:sexmale         0.238603002    0.49472054  0.4822986
primary:age            -0.010755256    0.01667258 -0.6450864
secondary:(Intercept)   0.351341730    0.77409402  0.4538748
secondary:sexmale      -0.293401655    0.47392024 -0.6190950
secondary:age          -0.004788914    0.01588194 -0.3015322

Differences between results based on synthetic and observed data:
                          Synthetic     Observed           Diff Std. coef diff
CI overlap
doctor:(Intercept)      0.167698852  0.287650834 -0.119951982       -0.153294
3  0.9608936
doctor:sexmale          0.406756817 -0.066070707  0.472827523        0.976230
6  0.7509570
doctor:age             -0.008294055 -0.005206560 -0.003087495       -0.190246
7  0.9514668
master:(Intercept)      0.925523568  0.373062189  0.552461379
0.7083414  0.8192973
master:sexmale         -0.103747550 -0.180835533  0.077087983        0.159407
5  0.9593341
master:age             -0.018554636 -0.005793053 -0.012761583       -0.785616
5  0.7995840
primary:(Intercept)     0.169180760  0.452962803 -0.283782044
-0.3579125  0.9086941
primary:sexmale         0.238603002 -0.238224469  0.476827471        0.963832
0  0.7541200
primary:age            -0.010755256 -0.009166959 -0.001588297       -0.095264
0  0.9756975
secondary:(Intercept)   0.351341730  0.132380101  0.218961629        0.282861
8  0.9278401
secondary:sexmale      -0.293401655 -0.235881218 -0.057520438       -0.121371
6  0.9690373
secondary:age          -0.004788914  0.002383676 -0.007172590
-0.4516194  0.8847889
```

```
Measures for 4 syntheses and 12 coefficients
Mean confidence interval overlap:  0.8884759
Mean absolute std. coef diff:  0.4371665

Mahalanobis distance ratio for lack-of-fit (target 1.0): 1.31
Lack-of-fit test: 15.70039; p-value 0.2053 for test that synthesis model is
compatible
with a chi-squared test with 12 degrees of freedom.
```

Confidence interval plot: You will see the graph shown in Figure 4-31.

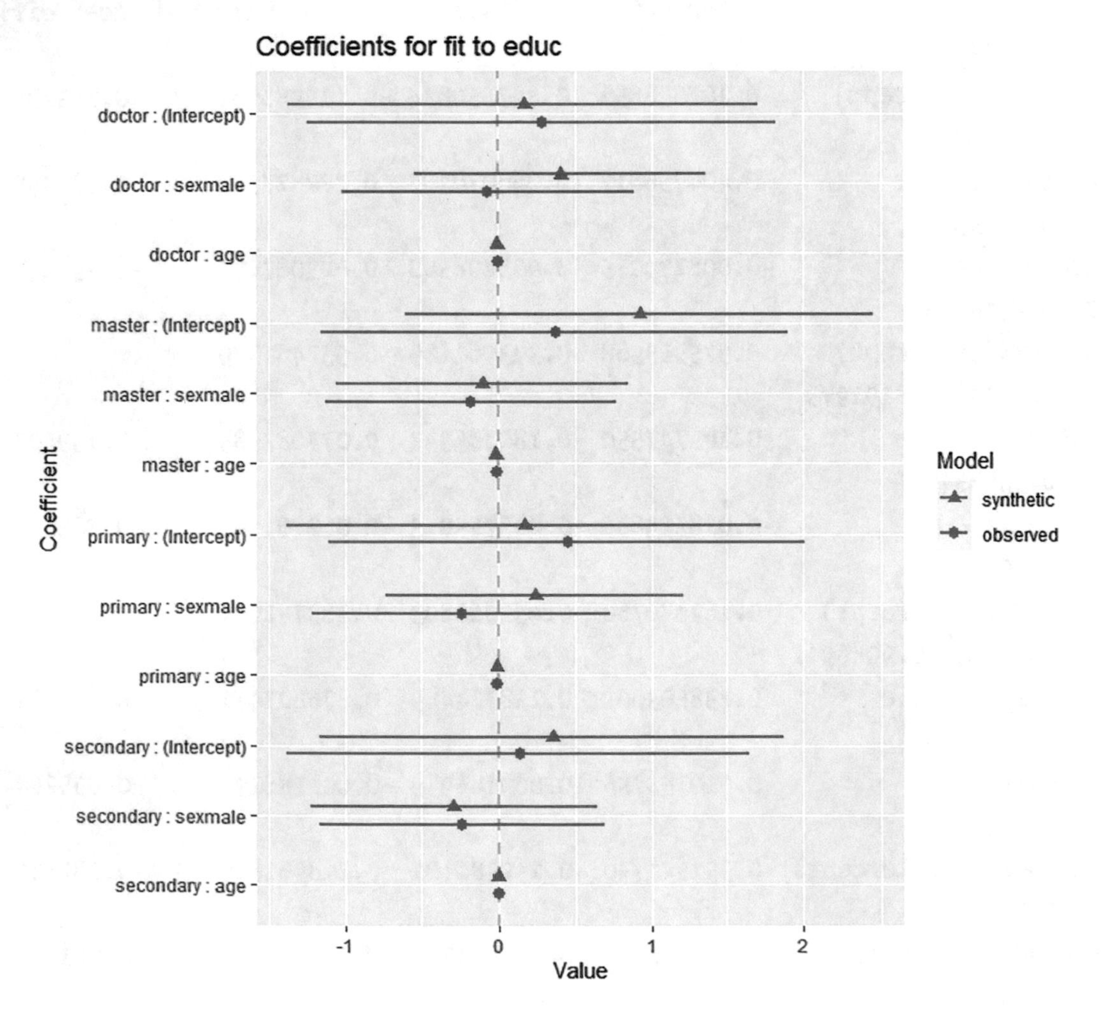

***Figure 4-31.** Confidence interval plot of the synthesized model for the educ variable*

It is seen that the properties of the betas, that is, the variables given in the above output, are similar. According to the expression "Mean confidence interval overlap: 0.8884759" expression given at the end of the above output, no statistically significant difference was found in the confidence interval comparisons. In addition, according to the expression "Lack-of-fit test: 15.70039; p-value 0.2053 for a test that synthesis model is compatible with a chi-squared test with 12 degrees of freedom" given at the end of the above output, it can be said that the synthesis model is compatible with real data.

In other words, it can be said that there will be no statistically significant difference between the synthesized data and the actual data. Thus it can be said that the synthesis was successful.

Copula

In statistics, it is necessary to look at the common distributions of these variables to capture the true relationship between random variables. The person correlation ρ, Spearman's rank ρ, and Kendall's τ methods simply provide a numerical value to measure. This numerical value is perceived by our mind and the real relationship is in the common distribution.

$$F(x_1, x_2, \ldots, x_n)$$

The copula function links the joint distribution of a set of variables to the marginal distributions of those variables. Copulas is very good at modeling and simulating correlated random variables. Using copulas, the correlation structure and marginal variables can be modeled separately. These models can often be useful. Because for some combinations of marginal variables, no function will generate the desired multivariate distribution. It is not easy to generate random samples with a distribution whose margins are Beta, Gamma and Student respectively. However, random samples from a multivariate normal distribution can be simply generated using R.

The copula function is mathematically represented by the symbol *C*. The *C* function is an application that maps a multivariate distribution to its univariate margins (marginal distributions). The two-dimensional copula function has the following two properties [7]:

$$C: \boldsymbol{I}^2 \rightarrow \boldsymbol{I}$$

$$1.\ \forall u,v \in \boldsymbol{I} \quad C(u,0)=0 \quad and \quad C(0,v)=0$$

$$C(u,1)=u \quad and \quad C(1,v)=v$$

$$2.\ \forall u_1,u_2,v_1,v_2 \in \boldsymbol{I},\ u_1 \le u_2 \quad and \quad v_1, \le v_2,$$

$$C(u_2,v_2)-C(u_2,v_1)-C(u_1,v_2)+C(u_1,v_1) \ge 0.$$

The $\boldsymbol{I}$ used in the above function is in the range [0, 1].

Let c denote the density of the copula. In this case, the copula density c is calculated as follows.

$$c = \frac{\partial C}{\partial F_1 \ldots \partial F_p}$$

Joint density is calculated as follows.

$$f(x) = c\left(F_1(x_1),\ldots,F_p(x_p)\right)\prod_{i=1}^{p} f_i(x_i)$$

The copula generator phi function can be used to create Archimedean copulas. Archimedean copulas are calculated as follows.

$$C(u,v) = \varphi^{[-1]}\left(\varphi(u)+\varphi(v)\right)$$

The Gumbel copula is a specific type of copula that is used to model the distribution of the failure time of a system. It is named after its creator, Harry Gumbel. Gumbel copula is created using the generator function below.

$$\varphi_\theta(t) = (-lnt)^\theta$$

When mathematical transformations are made, the following equation is obtained.

$$C_\theta(u,v) = exp\left[-\left((ln(u))\right)^\theta + \left(\left((-ln(v))\right)^\theta\right)^{1/\theta}\right], \theta \in [1,\infty]$$

In finance, it is often assumed that there are strong or weak correlations between multiple assets. There is always a correlation between the economic activities of human beings. These correlations are used to predict price and value movement. However, *t* is a big problem how to detect and measure these correlations is a big problem.

In finance, each asset can be assumed to have a return, such as the normal distribution $N(\mu, \sigma)$ or the student *t* distribution $t(n)$. But when it comes to investing, people prefer to diversify to minimize risk and keep several different instruments in their portfolios. Here, the weighted mean and variance can be used to estimate the parameter. However, the joint distribution of the portfolio is unknown. A common distribution can be produced using the known marginal distribution of assets. This way, it comes to the copula. Usually, Gaussian copula and *t* copula are used. During the financial crisis of 2008, MBS gathered most of its products. Gauss used the copula to estimate the distribution of MBS products and estimate the risk value.

t Copula

The *t* copula is a statistical tool that is used to measure the correlation between two variables. It that is used determine if there is a relationship between the two variables and to quantify that relationship. *t* copula is calculated as in the following equation.

$$C^t(u_1,\ldots,u_n;R_n,v)=t_{R_n,v}\left(t_v^{-1}(u_1),\ldots,t_v^{-1}(u_1)\right)$$

The density of *t* copula is defined by the following equation [8]:

$$C^t(u_1,\ldots,u_n;R_n,v)=\frac{f_{R_n,v}\left(t_v^{-1}(u_1),\ldots,t_v^{-1}(u_1)\right)}{\prod_{i=1}^{n} f_v\left(t_v^{-1}(u_i)\right)}$$

A *t* copula is a special case of the Gaussian copula and is often used when the variables are not normally distributed. The *t* copula is characterized by its heavy tails, which means that it is more likely than the Gaussian copula to generate extreme values. This makes it a good choice for modeling data with outliers.

In this example, the codes used to generate *t* copula are taken from `http://copula.r-forge.r-project.org/book/04_fitting.html`:

```
# Load the "copula" and "scatterplot3d" packages.
> library("copula")
```

```
> library("scatterplot3d")
>set.seed(300)
# In this method, an integer is selected because of pCopula().
>n<-6
# Generate of t copula.
>tCopula <- tCopula(iTau(tCopula(df =n), tau = 0.6), df =5)
# Random sample selection from t copula.
> X <- rCopula(1500, copula = tCopula)
```

Drawing the density graph of two-dimensional distributions. You will see the graph shown in Figure 4-32.

```
>wireframe2(tCopula, FUN = dCopula, delta = 0.050)
```

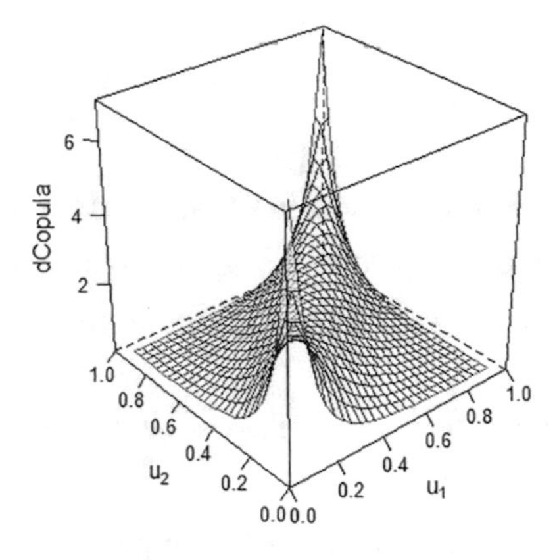

Figure 4-32. *Density plot of two-dimensional distributions*

Drawing the contour plot of the two-dimensional distribution. You will see the graph shown in Figure 4-33.

```
>contourplot2(tCopula, FUN = pCopula)
```

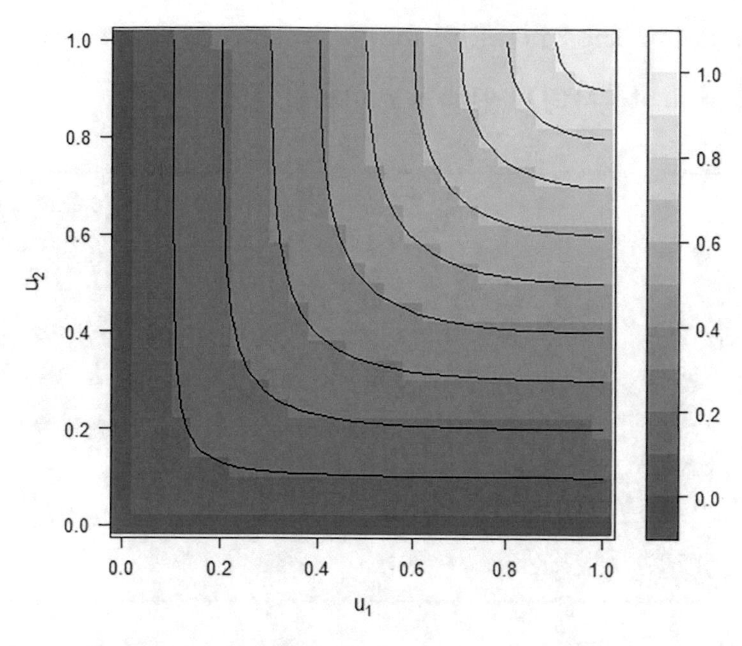

Figure 4-33. *Two-dimensional contour plot*

Contour density plot of the two-dimensional distribution. You will see the graph shown in Figure 4-34.

```
> contourplot2(tCopula, FUN = dCopula, n.grid = 40, cuts = 25)
```

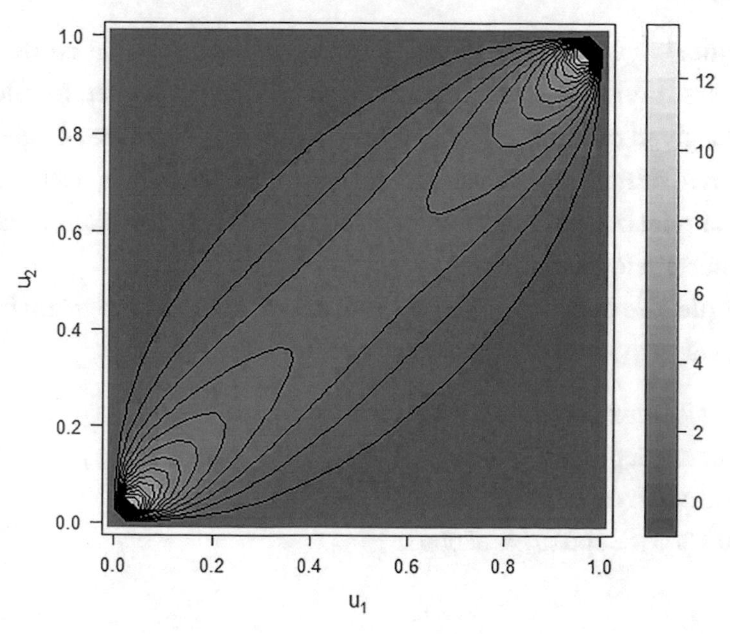

Figure 4-34. *Two-dimensional contour density graph*

Plotting the scatter plot. You will see the graph shown in Figure 4-35.

```
>plot(X, xlab = quote(X[1]), ylab = quote(X[2])
```

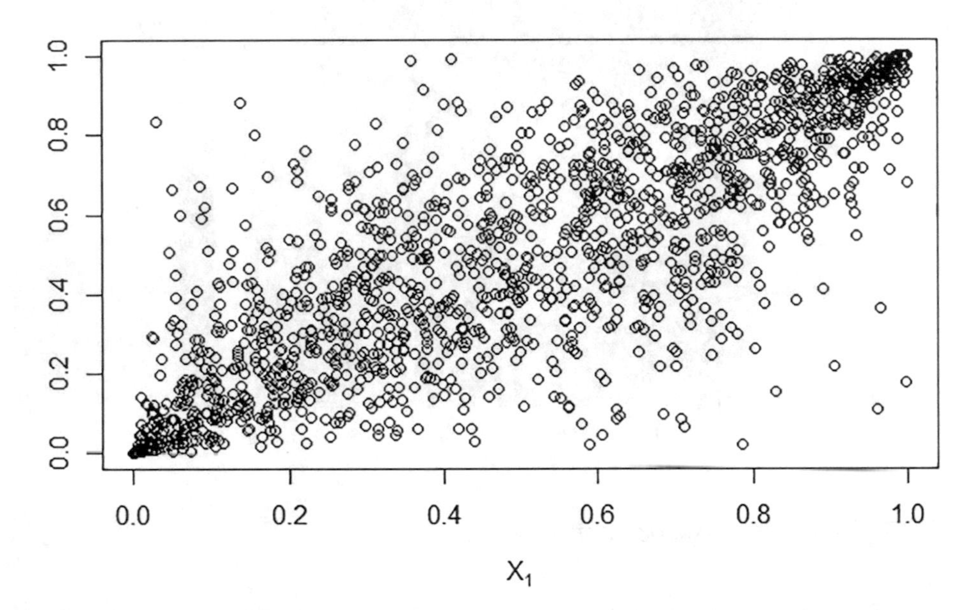

Figure 4-35. *Scatter plot of data*

Normal Copula

The normal copula is a function that describes the distribution of a random variable in terms of the distribution of two other random variables. It is used to calculate the probability that a given random variable will fall within a certain range given the distributions of two other random variables. The normal copula is particularly useful for modeling financial risk, where it can be used to calculate the likelihood of a portfolio experiencing a certain level of loss.

In this example, the codes used to generate normal copula are taken from `http://copula.r-forge.r-project.org/book/04_fitting.html`:

```
# Normal copula generation.
>nCopula <- normalCopula(iTau(normalCopula(), tau = 0.6))
# Random sampling from normal copula.
>X <- rCopula(1500, copula = nCopula)
```

Plotting the density of two-dimensional distributions. You will see the graph shown in Figure 4-36.

```
>wireframe2(nCopula, FUN = dCopula, delta = 0.050)
```

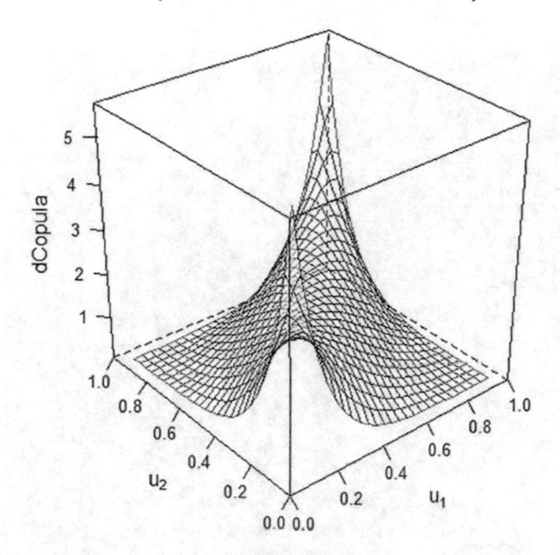

Figure 4-36. *Density plot of two-dimensional distributions*

Drawing the contour plot of the two-dimensional distribution. You will see the graph shown in Figure 4-37.

```
>contourplot2(nCopula, FUN = pCopula)
```

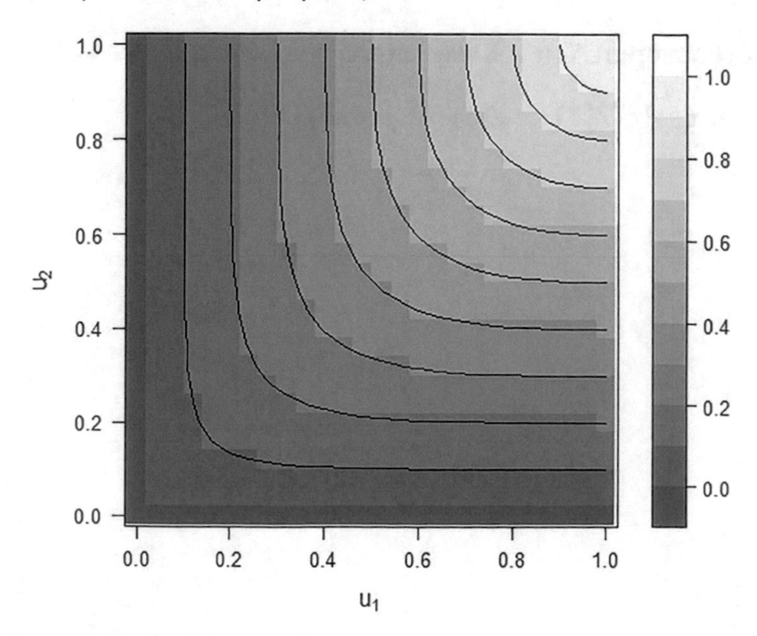

Figure 4-37. *Two-dimensional contour plot*

Contour density plot of the two-dimensional distribution. You will see the graph shown in Figure 4-38.

```
>contourplot2(nCopula, FUN = dCopula, n.grid = 44, cuts = 35, lwd = 1/4)
```

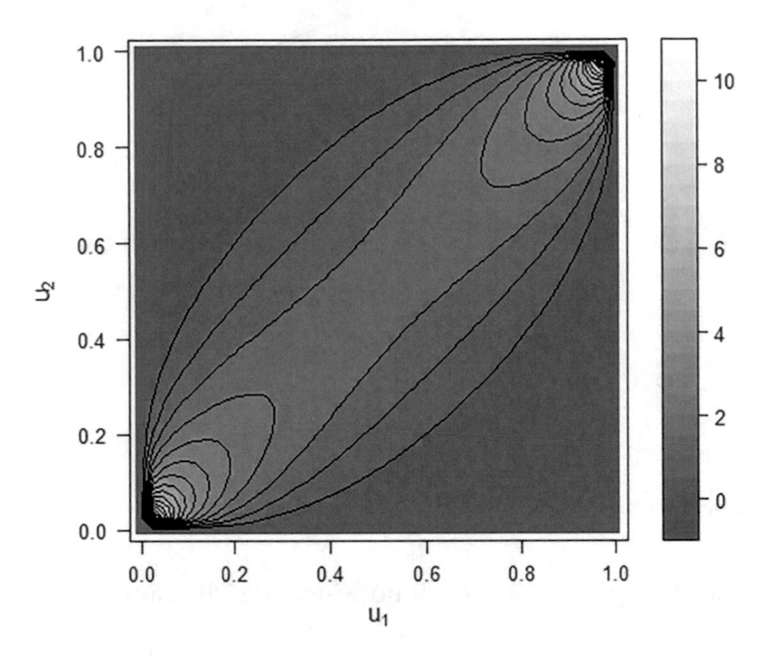

***Figure 4-38.** Two-dimensional contour density graph*

Plotting the scatter plot. You will see the graph shown in Figure 4-39.

```
>plot(X, xlab = quote(X[1]), ylab = quote(X[2]))
```

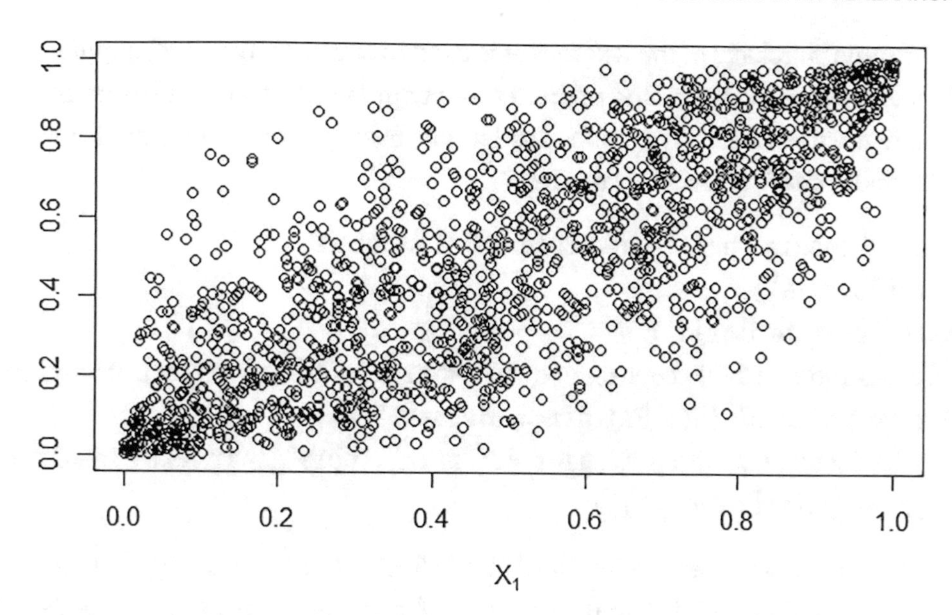

Figure 4-39. *Scatter plot of data*

Gaussian Copula

The Gaussian copula is a function that describes the dependence between two random variables. It is often used to model the joint distribution of two variables, which may be normal or not normal. The Gaussian (Normal) copula is calculated as follows.

$$C(u,v;\theta)=\Phi_G\left(\Phi^{-1}(u),\Phi^{-1}(v);\theta\right)$$
$$=\int_{-\infty}^{\Phi^{-1}(u)}\int_{-\infty}^{\Phi^{-1}(v)}\frac{1}{2\pi\left(1-\theta^2\right)^{1/2}}\times\left\{\frac{-\left(x^2-2\theta xy+y^2\right)}{2\left(1-\theta^2\right)}\right\}dxdy$$

The $\Phi^{-1}(.)$ in the above equation is the inverse function of the standard normal distribution ($\Phi(.)$). θ is the linear correlation coefficient between $\Phi^{-1}(u)$ and $\Phi^{-1}(v)$.

The Gaussian copula density function is a mathematical function that describes the distribution of a set of jointly Gaussian random variables. This function is calculated as follows.

$$c(x)=\frac{1}{|R|^{1/2}}exp\left\{-\frac{1}{2}u'\left(\mathrm{R}^{-1}-I\right)u\right\}$$

The "copula" package of the R software is used to statistically model copulas. Two objects are created here. One of them is the copula object, and the other is the multivariate normal object. The codes used in this example are taken from here `https://rpubs.com/lance4869/copula`:

```
# Load the "copula" and "scatterplot3d" libraries.
> library("copula")
> library("scatterplot3d")
# Two-dimensional (2-D) copula and two-variable normal object creation.
> mycop<-normalCopula(c(0.82),dim=2,dispstr="ex")
> mymvd<-mvdc(copula=mycop,margins =c("norm","norm"),paramMargins=list(list
(mean=0,sd=1),list(mean=1,2)))
```

The correlation between the two variables is 0.82. "dispstr" is the type of correlation matrix. "ex" is the power of 0.82 to the distance of the index x_i and x_j, which is $|i - j|$ is expressed as.

```
# Generating 1500 random numbers from a multivariate distribution.
> r<-rMvdc(1500,mymvd)
# Calculation of density.
> density<-dMvdc(r,mymvd)
# Calculation of the cumulative distribution.
> distance<-pMvdc(r,mymvd)
```

Visualization of the density graph in three-dimensional space. You will see the graph shown in Figure 4-40.

```
> x<-r[,1]
> y<-r[,2]
> scatterplot3d(x,y,density,highlight.3d = T)
```

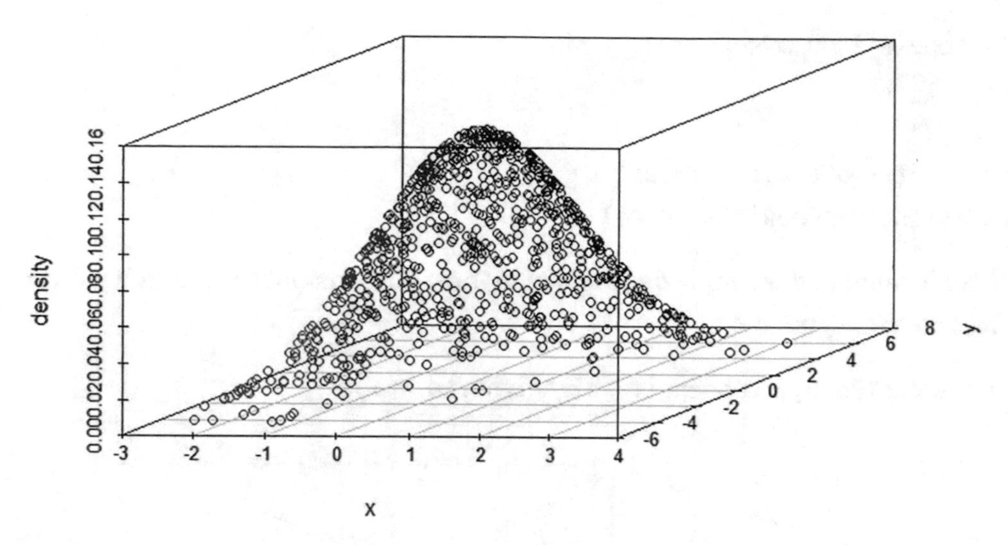

Figure 4-40. *Density graph in three-dimensional space*

Visualization of the distance graph in three-dimensional space. You will see the graph shown in Figure 4-41.

```
> scatterplot3d(x,y,distance,highlight.3d = T)
```

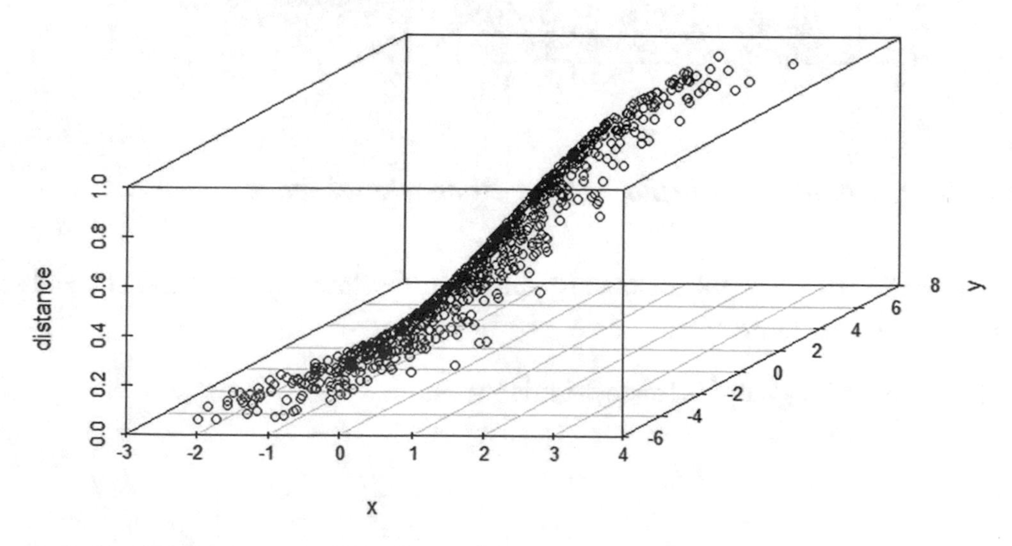

Figure 4-41. *Distance graph in three-dimensional space*

```
# It is also possible to visualize the copula function. The copula function
is visualized as follows.
```

```
> w<-rCopula(1500,mycop)
> x<-w[,1]
> y<-w[,2]
> copdensity<-dCopula(w,mycop)
> copdistance<-pCopula(w,mycop)
```

Visualization of the copula density plot in three-dimensional space. You will see the graph shown in Figure 4-42.

```
> scatterplot3d(x,y,copdensity,highlight.3d = T)
```

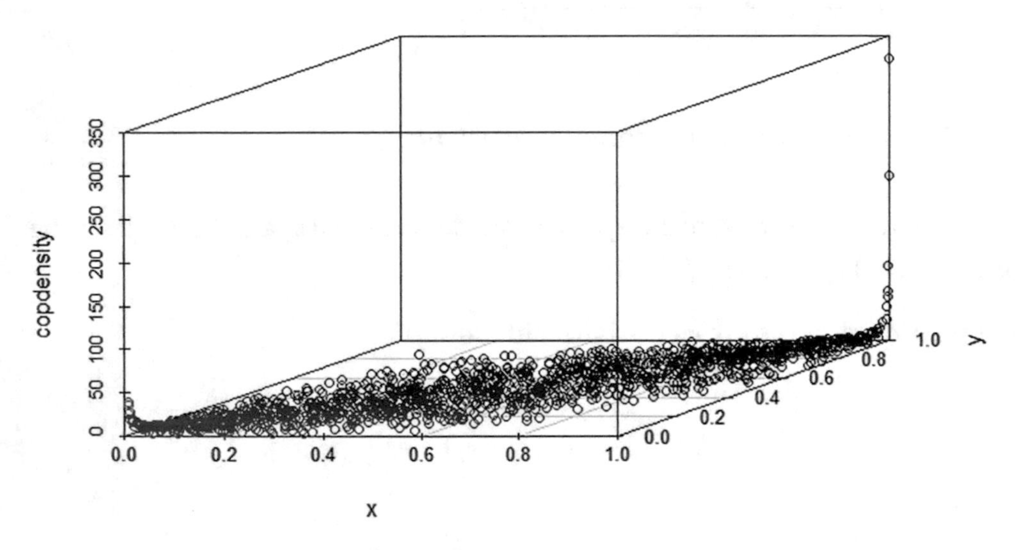

Figure 4-42. *Copula density plot in three-dimensional space*

Visualization of the copula distance graph in three-dimensional space. You will see the graph shown in Figure 4-43.

```
> scatterplot3d(x,y,copdistance,highlight.3d = T)
```

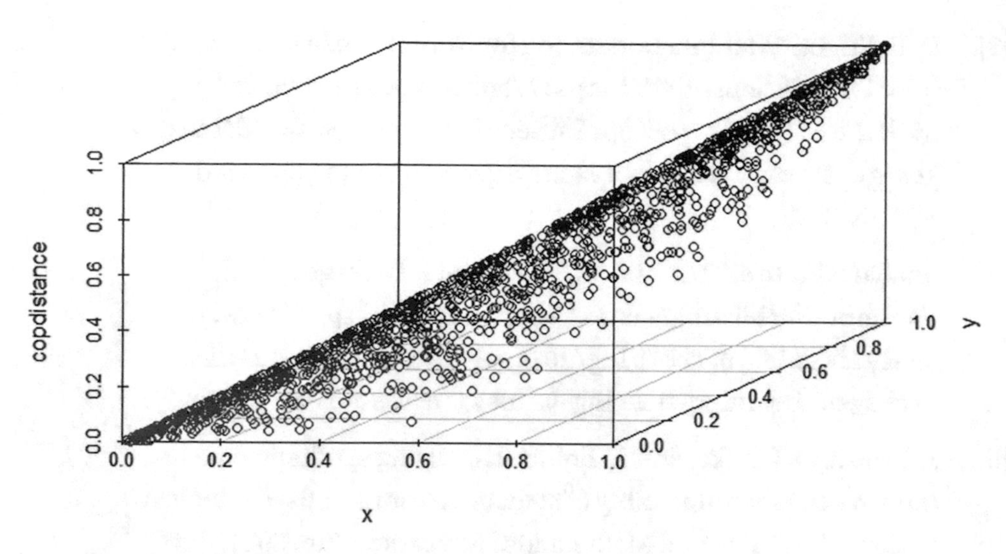

Figure 4-43. *Copula distance graph in three-dimensional space*

Summary

In this chapter, you learned how to generate synthetic data in R using different packages and functions. You learned how to create a value vector from a known univariate distribution. You also learned how to generate synthetic data from a multivariate categorical variable. You learned how to generate synthetic data from a multivariate distribution with correlation. You learned how to generate an artificial neural network using the nnet package in R. You also learned about augmented data, and image augmentation using the torch package. Also, you learned how to generate synthetic data using the mice, conjurer, and synthpop package in R. Finally, you learned how to generate synthetic data using the copula package in R.

Next, we'll being delving into Synthetic Data Generation with Python.

References

[1]. C. Shorten and T. M. Khoshgoftaar, "A survey on Image Data Augmentation for Deep Learning," Journal of Big Data, vol. 6, no. 1, p. 60, Dec. 2019, doi: 10.1186/s40537-019-0197-0.spiepr Par165

[2]. G. Andrews, "What Is Synthetic Data?," NVIDIA Blogs, Jun. 08, 2021. `https://blogs.nvidia.com/blog/2021/06/08/what-is-synthetic-data/` (accessed Apr. 15, 2022).

[3]. D. B. Rubin, "Multiple Imputation for Nonresponse in Surveys," John Wiley & Sons, 2004. https://books.google.com.tr/books?id=bQBtw6rx_mUC&printsec=frontcover&hl=tr&source=gbs_ge_summary_r&cad=0#v=onepage&q&f=false (accessed Apr. 15, 2022).

[4]. AnalyticsVidhya, "Tutorial on 5 Powerful R Packages Used For Imputing Missing Values," Mar. 04, 2016. https://www.analyticsvidhya.com/blog/2016/03/tutorial-powerful-packages-imputing-missing-values/ (accessed Apr. 15, 2022).

[5]. S. Grund, O. Lüdtke, and A. Robitzsch, "Using synthetic data to improve the reproducibility of statistical results in psychological research," Science and Mathematics Education, Jun. 2021, doi: 10.31234/OSF.IO/D7ZWJ.

[6]. T. B. Volker and G. Vink, "Anonymiced Shareable Data: Using mice to Create and Analyze Multiply Imputed Synthetic Datasets," Psych, vol. 3, no. 4, pp. 703–716, Nov. 2021, doi: 10.3390/psych3040045.

[7]. R. B. Nelsen, An Introduction to Copulas, Springer. New York: Springer, 2006.

[8]. F. Ielpo and C. Merhy, "Engineering Investment Process: Making Value Creation Repeatable", Elsevier, 2017.

CHAPTER 5

Synthetic Data Generation with Python

In this chapter, we will explore how to generate synthetic data for regression, classification, and clustering problems using Python. First, we will discuss how to generate synthetic data from a known distribution. Next, we will apply Gaussian noise to a regression model. Then, we will discuss how to generate synthetic data for classification and clustering problems using the Friedman functions and symbolic regression. Finally, we will generate synthetic data for tabular data using GANs.

Data Generation with Know Distribution

The synthetic data was generated by a computer program that followed a specific set of rules or conditions. The rules were designed to create data that would have the same statistical properties as the original data set but would not contain any actual information about the original data set. The generated data set would be used to test various data analysis methods without the risk of revealing any confidential information.

We can generate synthetic data that conforms to a known distribution in several ways. One way is to use a parametric model that specifies the functional form of the underlying distribution, such as a normal distribution. We can then generate data points from this parametric model that will have the same distribution as the underlying data.

Another way to generate synthetic data with a known distribution is to use a non-parametric model, such as kernel density estimation. This approach does not require us to specify the functional form of the underlying distribution but rather estimates it from the data itself. This can be useful when the underlying distribution is not well-understood, or when we don't have enough data to accurately specify a parametric model.

N. Gürsakal et al., *Synthetic Data for Deep Learning*, https://doi.org/10.1007/978-1-4842-8587-9_5

Once we have generated our synthetic data, we can then use it for any purpose that we would use real data, such as training machine learning models or performing statistical analysis.

All samples will generate based on the distribution with given parameters. We have different distributions, such that Normal, Uniform, Triangular, Binomial, Chi-square, Laplace, and of them parameters.

```
import numpy as np # calling the numpy library
import matplotlib.colors # calling the numpy matplotlib.colors library
import matplotlib.pyplot as plt# calling the matplotlib.pyplot library
d_list = ['normal','uniform','triangular','binomial','chisquare','laplace']
# define which distribution to be used
p_list = ['0,1','-1,1','-3,0,8','10,0.5','2','5,4'] # each distribution
parameters
c_list = ['red','limegreen','gold','purple','blue','magenta'] # Each
graph colors
fig, axs = plt.subplots(nrows=2, ncols=3, dpi=800,figsize=(7,6))
k=0
for i in range(2):
    for j in range(3):
        datas=eval("np.random."+d_list[k]+"("+p_list[k]+",5000)")
        axs[i][j].hist(datas, bins=50,color=c_list[k])
        axs[i][j].set_title(d_list[k]+" dist( "+p_list[k]+")",size=8)
        k+=1
plt.suptitle('Samples from Different Distributions',fontsize=15)
fig.savefig("Dist Histogram.png")
plt.close(fig)
```

You will see the graph shown in Figure 5-1.

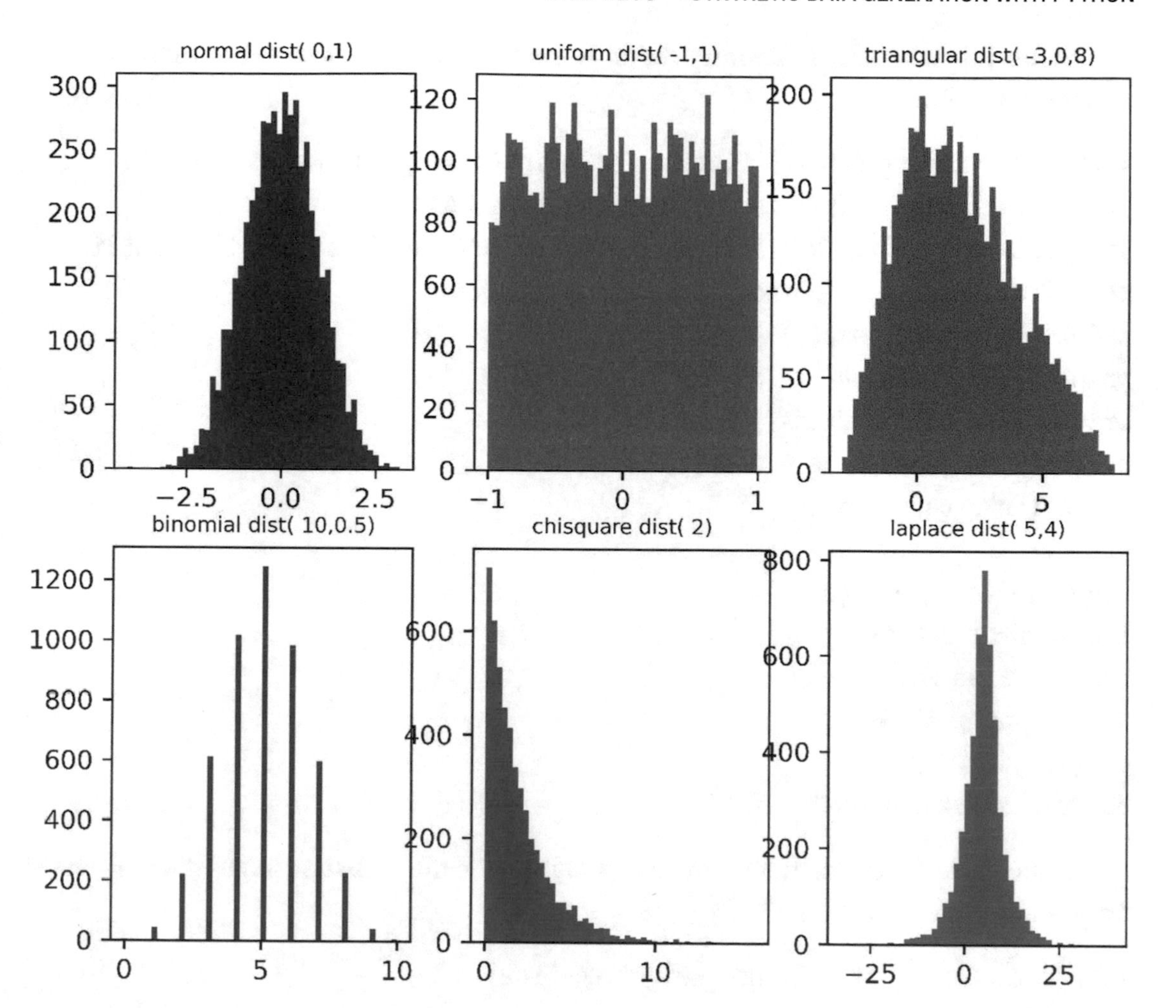

Figure 5-1. *Samples from different distributions*

Also, sometimes we need to have fake text data to fill out complete any database. Faker is a Python library that generates fake text data.

```
from faker import Faker # calling the library

    fake_data=Faker()
```

Default is the American database, but you can change this database to any nationality.

All our fake_data comes from the Faker Class, so in this class there is too much different information.

To create a Turkish data profile, we can easily determine the "tr_TR" nationality in the Faker() function.

Name, job, address, phone number, and email data can be easily call with the Faker () function.

```
from faker import Faker # Starting with installing the Faker library
fake_data=Faker()# This is the American data base
fake_data=Faker("tr_TR") # Created Turkish databased, so all the results
come from Turkish databsed
print (fake_data.name())
print (fake_data.job())
print (fake_data.address())
print (fake_data.phone_number())
print (fake_data.email())
```

Output:

```
Deha Çamurcuoğlu Akça
Laboratuvar işçisi
82750 Arslan Mews
Akgüneşport, MI 04351
+90(526)8817713
nurmelek56@example.net
```

In the Faker function, there is the use_weighting features that arranges the real-world frequency of the values.

```
fake_data=Faker("tr_TR", use_weighting=True)
for i in range(10):
    print (fake_data.name())
```

Output:

```
Bayan Zubeyde Şemsinisa Soylu
Ulu Göksev Durmuş
Uçan Eraslan
Ozansü Ertün Demir
Sanur Döner Şensoy
Tekiner Akçay
Nalân Gülen
Vildane Duran
Dr. Aydınbey Sunel Türk
Dr. Hasbek İntihap Akçay
```

We can easily compare the results by using the use_weighting function or not.

Data with Date information

Define a fake birth date, or date.

```
print (fake_data.date_of_birth())
print (fake_data.date())Output:
1949-12-22
1995-04-09
```

Data with Internet information

```
print (fake_data.hostname())
print (fake_data.ipv4())
Output:
db-88.turk.com
181.228.175.188
```

A more complex and comprehensive example

We will create a new csv document that covers patients' records for the international hospital with the help of the Faker () function.

```
from faker import Faker # Calling Faker library
import pandas as pnd # Calling panda library
import numpy as np # Calling numpy library
fake_data=Faker(["en_US","fr_FR","tr_TR"], use_weighting=True) # More than
one different language database is defined
symptoms_list=["anxiety","depression","back_pain","diarrhea","fever",
"dizzy","cough","apnea"] # List of specified symptoms
patients={}
for k in range(0,1000):# 1000 patients are created in total
    patients[k]={} # Creating patient list and id, name, address, phone
number, date of birth is assigned to each patient.
```

```
    patients[k]['id']=k+1
    patients[k]['name']=fake_data.name()
    patients[k]['address']=fake_data.address()
    patients[k]['phone_number']=fake_data.phone_number()
    patients[k]['Date of Birth']=fake_data.date()    patients[k]
['symptoms']=np.random.choice(symptoms_list) # Random assignment of
symptoms from a list of specified symptoms
data_frame=pnd.DataFrame(patients).T
print (data_frame)
data_frame.to_csv("Patinets_data.csv", index=False) # Generated dataset is
saved in cvs file
```

Output:

```
       id                name  ... Date of Birth  symptoms
0       1        Chad Webster  ...    1990-03-24  diarrhea
1       2        Amanda Watts  ...    2005-01-19     dizzy
2       3        Craig Decker  ...    2003-08-29     apnea
3       4        Aurore Boyer  ...    1989-09-03     fever
4       5       Lisa Edwards  ...    2016-09-25     cough
..    ...                 ...  ...           ...       ...
995   996       April Pollard  ...   2013-03-24   anxiety
996   997         Yazgül Çetin  ...    1996-07-02     apnea
997   998  Rémy Julien-Texier  ...    2015-06-24  diarrhea
998   999      Robin Blanchard  ...    2016-08-03     apnea
999  1000    Martine Schmitt  ...    1999-10-17     apnea
[1000 rows x 6 columns]
```

Synthetic Data Generation in Regression Problem

In regression analysis, synthetic data generation is the process of creating new data that is similar to existing data but is not identical to it. This new data can be used to train a model or to test a hypothesis. There are many reasons why you might want to generate synthetic data. For example, you might want to create a dataset that is larger than what you have available, or you might want to create a dataset with more diverse data than what you have available. Synthetic data can also be used to fill in missing values in a dataset. There are a variety of methods that can be used to generate synthetic data. Some methods are more complex than others, and some will be more accurate than others. The method that you choose will depend on the type of data that you are working with and the objectives that you are trying to achieve.

The sklearn.dataset package in the Python library generates a random regression data set. The make_regression() function in this package consists of n_samples, n_features, and a target variable mainly. n_sample mentions the number of samples to generate, n_features create independent *x* variables as much as we want, and target variables play roles as our dependent variables.

```
from faker import Faker
import pandas as pnd
import numpy as np
import matplotlib.pyplot as plt # Required for drawing graphics
from sklearn.datasets import make_regression # Required for generating a
random regression data set

c_mapp=plt.cm.get_cmap("YlGnBu") # Color of the graph
data_reg=make_regression(n_samples=1000, n_features=2, noise=0.0) # 1000
samples are created with two features
print (data_reg[0])
data_frame1= pnd.DataFrame(data_reg[0],columns=['x'+str(i) for i in
range(1,3)])
data_frame1['y'] = data_reg[1]
print (data_frame1)
```

Output:

```
           x1        x2           y
0    0.451775  1.126216  105.637343
1   -1.153118  2.894067  173.656334
2   -0.963961  1.377513   65.112484
3   -1.460770 -1.414778 -170.456499
4   -1.225991 -1.876402 -196.006796
..        ...       ...         ...
995 -0.629005 -1.539042 -144.851247
996 -1.281187 -1.064484 -135.948385
997  0.438089  0.611139   65.478204
998  1.513318 -1.033047  -15.423906
999 -0.770491 -0.248904  -51.691303
```

```
fig, axs = plt.subplots(figsize=(9,5)) # Define figure size
axs.scatter(data_frame1.x1,data_frame1.x2, cmap=c_mapp,c=data_
frame1.y,vmin=min(data_frame1.y), vmax=max(data_frame1.y))
axs.set_title('noise=0') # set title
plt.show()
```

Output:

You will see the graph shown in Figure 5-2.

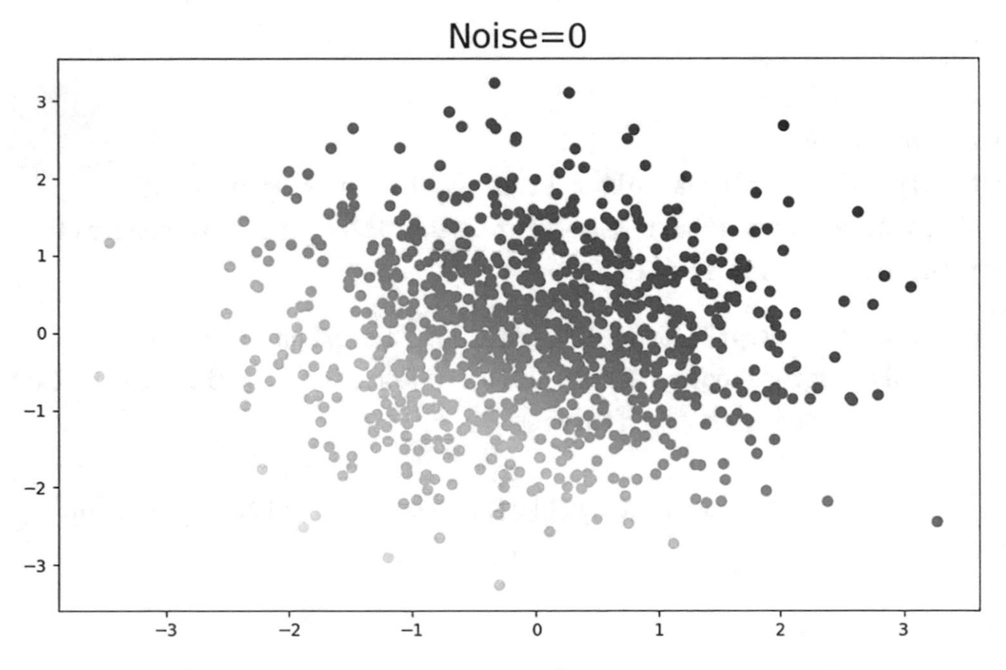

***Figure 5-2.** Random Regression Data set without noise*

We can also make a regression plot for each x data set with a fitted regression line.

```
data_reg=make_regression(n_samples=30, n_features=2,
              noise=0.0) ) # 30 samples are created with two features
data_frame1= pnd.DataFrame(data_reg[0],columns=['x'+str(i) for i in
range(1,3)])
data_frame1['y'] = data_reg[1]
```

```
fig, ax = plt.subplots(2,figsize=(9,5)) # Two feature will be displayed on
two figures
reg_fit_x1=np.polyfit(data_frame1.x1,data_frame1.y,1) # First feature
fit_function1=np.poly1d(reg_fit_x1) # Regression fit line
reg_fit_x2=np.polyfit(data_frame1.x2,data_frame1.y,1) # Second feature
fit_function2=np.poly1d(reg_fit_x2) # Regression fit line
# Adjusting the figure's properties, such that color, size
ax[0].scatter(data_frame1.x1,data_frame1.y, s=100,c="red",
edgecolor="black")
ax[0].plot(data_frame1.x1,fit_function1(data_frame1.x1),':b', lw=2)
ax[1].scatter(data_frame1.x2,data_frame1.y, s=100,c="red",
edgecolor="black")
ax[1].plot(data_frame1.x2,fit_function2(data_frame1.x2), ':b', lw=2)
plt.show()
```

Output:

You will see the graph shown in Figure 5-3.

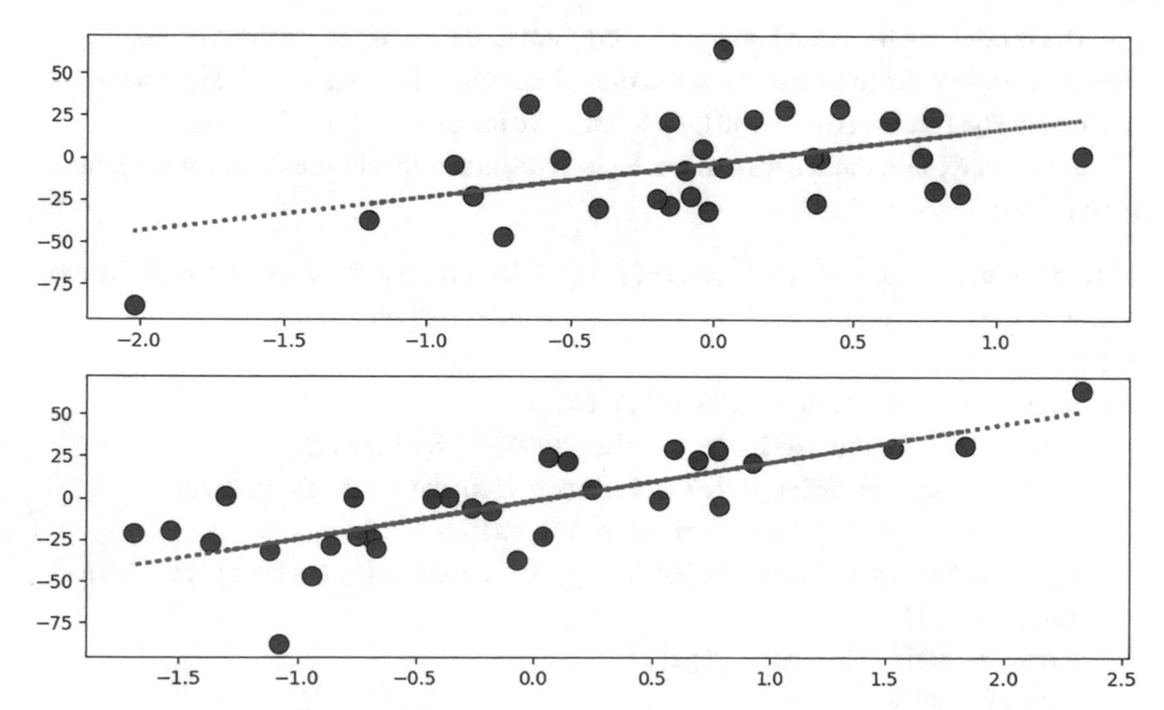

***Figure 5-3.** Fitted regression line*

Gaussian Noise Apply to Regression Model

Regression models are used to predict continuous values. They are commonly used to predict things like prices or demand.

Adding noise to data is a common way to simulate real-world conditions. It can help to make sure that a model is not overfitting to the training data.

In this example, we will add Gaussian noise to synthetic data and see how it affects a regression model. We will generate two data sets. One will be used to train the model and the other will be used to test the model.

The training data will have noise added to it. The test data will not have any noise added. This will allow us to see how well the model can generalize to new data.

We will use a simple linear regression model for this example. The model will have one input variable and one output variable. The input variable will be the *x-value of the data point,* and the output variable will be the *y*-value of the data point.

We will add Gaussian noise with a mean of 0 and a standard deviation of 1 to the training data. This will create a dataset with *x*-values that are randomly distributed around the *y*-value.

The model will be trained on the training data and then tested on the test data. We will compare the predicted values to the actual values to see how well the model performs. The Gaussian noise will help to improve the accuracy of our model.

In the regression model, we can implement the noise effect to see how created data is changing.

```
fig, ax = plt.subplots(3,2,figsize=(9,5)) # Generating figures for 6 different
noise values consisting of 3 rows and 2 columns for each noise value
a=231
for noise_data in [1,10,50,100,500,1000,]:
   data_reg2=make_regression(n_samples=1000, n_features=2,
            noise= noise_data) # A regression data set is created
            according to different noise values.
    data_frame1= pnd.DataFrame(data_reg2[0],columns=['x'+str(i) for i in
    range(1,3)])
    data_frame1['y'] = data_reg2[1]
    plt.subplot(a)
    plt.scatter(data_frame1.x1,data_frame1.x2, cmap=c_mapp,c=data_
    frame1.y,vmin=min(data_frame1.y), vmax=max(data_frame1.y))
```

```
    plt.title('Noise='+ str(noise_data), size=10)
    a+=1
plt.show()
```

You will see the graph shown in Figure 5-4.

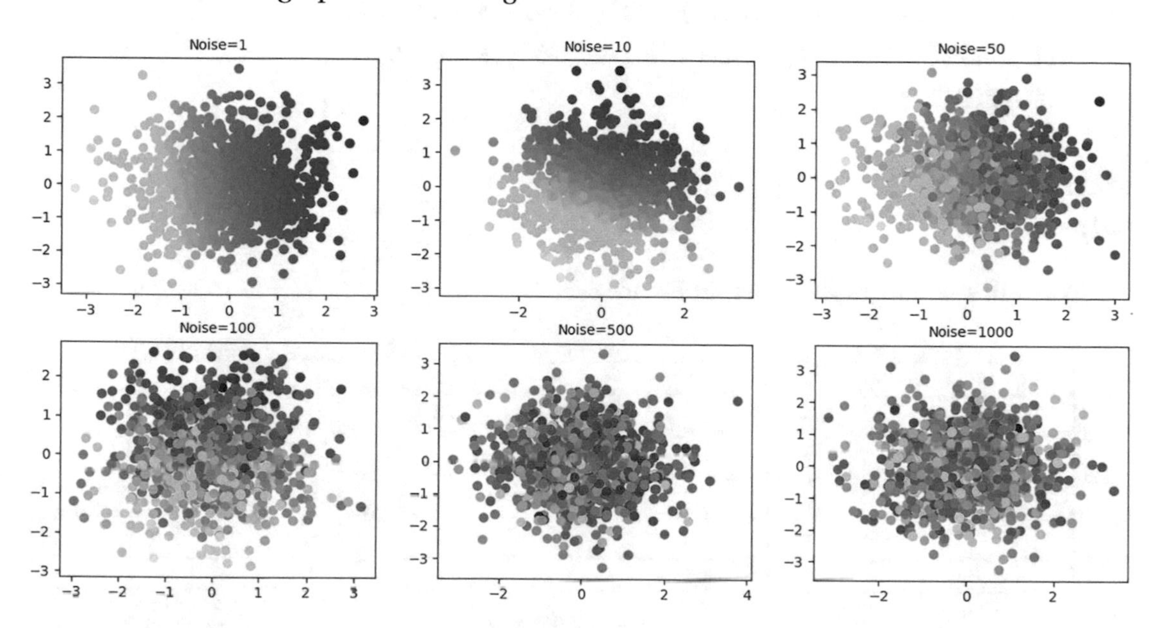

Figure 5-4. *Regression datas Data set with Gaussian Noise*

The regression model of the x dataset was produced with 6 variables, with the noise variable being 500 graphs below,

```
fig, ax = plt.subplots(3,2,figsize=(9,5)) # Generating figures for 6
features values consisting of 3 rows and 2 columns
a=231
noise_data=500
data_reg3=make_regression(n_samples=30, n_features=6,
               noise=noise_data) # 30 samples are created with 6 features
data_frame1= pnd.DataFrame(data_reg3[0],columns=['x'+str(k) for k in
range(1,7)])
data_frame1['y'] = data_reg3[1]
for i in range(6):
    reg_fit=np.polyfit(data_frame1[data_frame1.columns[i]],data_frame1.y,1)
    fit_function1=np.poly1d(reg_fit) # Regression fit line
```

```
    plt.subplot(a)
    plt.scatter(data_frame1[data_frame1.columns[i]],data_frame1.y,
s=100,c="red", edgecolor="black") # Scatter plot
    plt.plot(data_frame1[data_frame1.columns[i]],fit_function1(data_
frame1[data_frame1.columns[i]]),':b', lw=2) # Line plot
    plt.title('X'+str(i)+" with Noise=500", size=10)
    plt.grid(True)
    a+=1
plt.show()
```

Output:

You will see the graph shown in Figure 5-5.

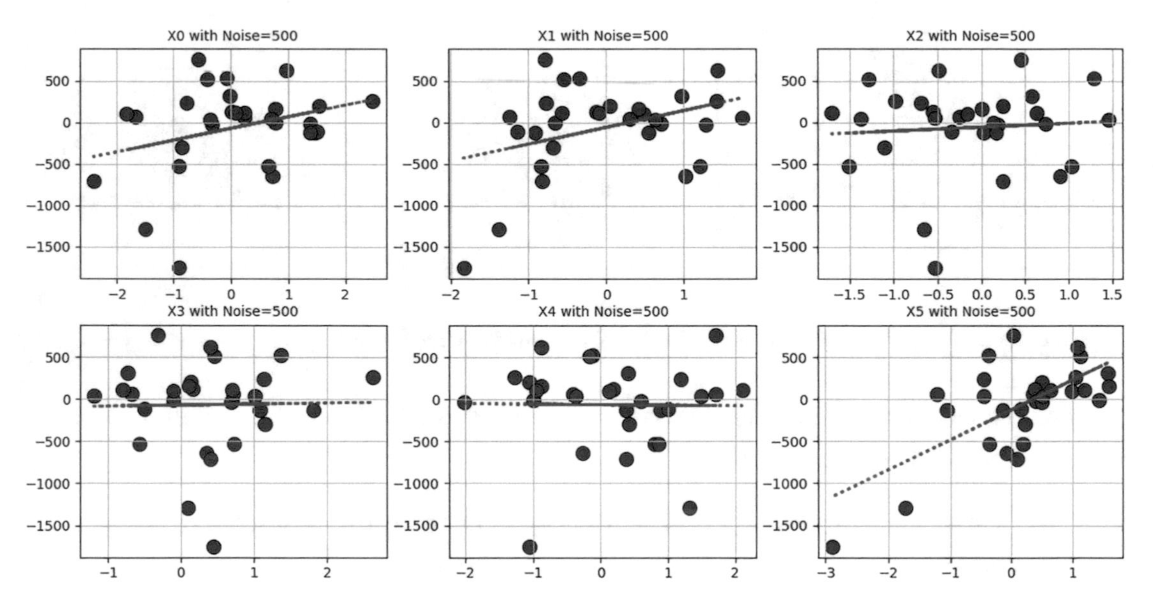

Figure 5-5. *Fitted regression line with same Gaussian Noise*

We can plot the data set with different degrees of noise

```
fig, ax = plt.subplots(3,2,figsize=(9,5))
data_frame=pnd.DataFrame(data=np.zeros((30,1)))
a=231
noise_data=[1,10,50,100,500,1000,] # Noise values
for i in range(6):
```

```
    data_reg4=make_regression(n_samples=30, n_features=1,
                              noise=noise_data[i]) # 30 samples are created with
                              1 features and certain noise values
    data_frame["x"+str(i+1)]=data_reg4[0]
    data_frame["y"+str(i+1)]=data_reg4[1]
for i in range(6):
    reg_fit=np.polyfit(data_frame["x"+str(i+1)],data_frame["y"+str(i+1)],1)
    fit_function1=np.poly1d(reg_fit) # Regression fit line
    plt.subplot(a)
    plt.scatter(data_frame["x"+str(i+1)],data_frame["y"+str(i+1)],
    s=100,c="red", edgecolor="black") # Scatter plot
    plt.plot(data_frame["x"+str(i+1)],fit_function1(data_
    frame["x"+str(i+1)]),':b', lw=2) # Regression line
    plt.title('Noise='+ str(noise_data[i]), size=10)
    plt.grid(True)
    a+=1
plt.show() #
```

Output:

You will see the graph shown in Figure 5-6.

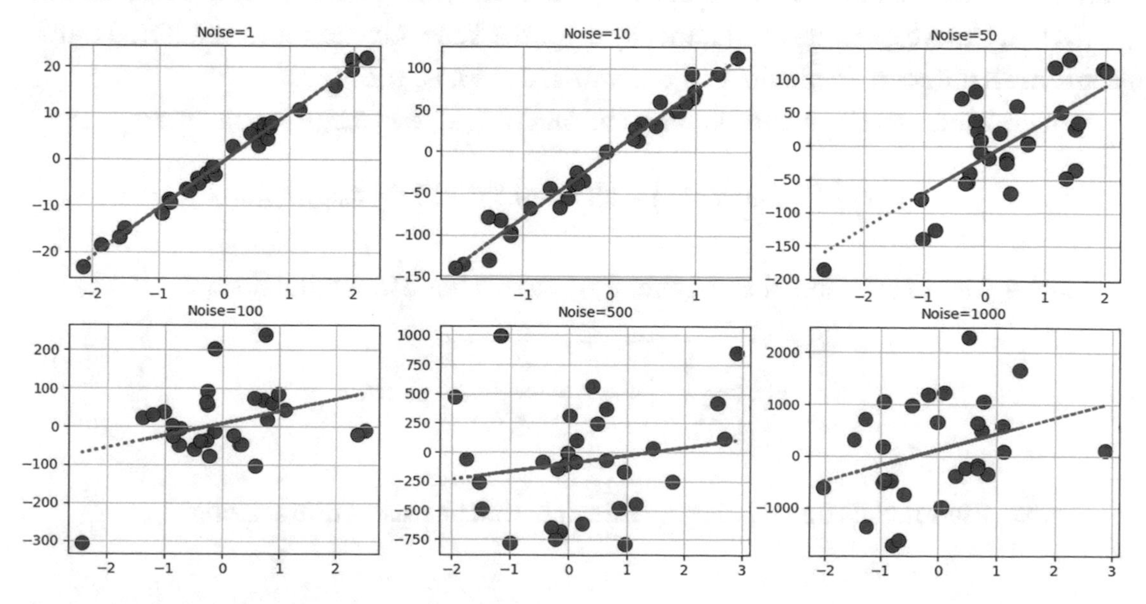

***Figure 5-6.** Fitted regression line with different Gaussian Noise*

Friedman Functions and Symbolic Regression

Friedman functions are a type of mathematical function used in statistics and regression analysis. They are named after statistician Milton Friedman, who introduced them in his work on nonlinear regression.

Friedman functions are used to model data that are not linearly related. That is, the data cannot be modeled using a straight line. Instead, the data are modeled using a curved line. Friedman functions are often used in symbolic regression. Symbolic regression is a type of machine learning that automatically finds mathematical equations that best describe a data set. Friedman functions are particularly well suited for symbolic regression because they can model data that are not linearly related. This is important because many real-world data sets are not linearly related.

The use of Friedman functions in symbolic regression is very effective in synthetic data sets. Friedman functions are also used in other types of machine learning, such as support- vector machines and artificial neural networks. Also, the use of Friedman functions in machine learning is an active area of research.

Friedman functions are particularly well suited for data that are "noisy" or have many outliers. This is because they are less sensitive to small changes in the data than other functions, such as polynomials. Friedman functions can be used to model a wide variety of data, including financial data, scientific data, and even data from social media.

When can we generate the data set by using the know function such that Friedman? There are 3 different Friedman data generation formulas. Those are:

make_friedman1() function is generate data with at least 5 input parameters.

$$y(x) = 10 * \sin(\pi x_0 x_1) + 20(x_2 - 0.5)^2 + 10x_3 + 5x_4 + noise$$

make_friedman2() function is generate data that has 4 input dimensions.

$$y(x) = \sqrt{\left(x_0^2 + x_1 x_2 - \frac{1}{(x_1 x_3)^2}\right)} + noise$$

make_friedman3() function is generate data that has also 4 dimensions.

$$y(x) = \arctan\left(\frac{x_1 x_2 - \frac{1}{x_1 x_3}}{x_0}\right) + noise$$

Data set has been created by using these functions.

```
from faker import Faker # Required libraries faker, pandas, numpy,
matplotlib.pyplot, sklearn.datasets
import pandas as pnd
import numpy as np
import matplotlib.pyplot as plt
import sklearn.datasets as skl_dataset
c_map=plt.cm.get_cmap("YlGnBu") # Color of the figure
x_variables,y = skl_dataset.make_friedman1(n_samples=1500,n_features=6,
noise=0.0) # 1500 samples are created with 6 features and without any noise

data_frame1=pnd.DataFrame(x_variables,columns=['x'+str(i) for i in
range(1,7)])
data_frame1['y'] = y
print (data_frame1)
```

Output:

	x1	x2	x3	x4	x5	x6	y
0	0.548814	0.715189	0.602763	0.544883	0.423655	0.645894	17.213492
1	0.437587	0.891773	0.963663	0.383442	0.791725	0.528895	21.503940
2	0.568045	0.925597	0.071036	0.087129	0.020218	0.832620	14.619807
3	0.778157	0.870012	0.978618	0.799159	0.461479	0.780529	23.373800
4	0.118274	0.639921	0.143353	0.944669	0.521848	0.414662	16.955281
...	...	...	...	...	...	...	...
1495	0.672339	0.941159	0.690350	0.559549	0.157171	0.921106	16.248550
1496	0.996720	0.842478	0.093479	0.111422	0.364915	0.696023	11.069381
1497	0.826904	0.180815	0.625240	0.159566	0.112104	0.470972	6.996266
1498	0.861044	0.627706	0.681544	0.393166	0.266880	0.932096	15.844463
1499	0.411408	0.513405	0.072222	0.069151	0.758064	0.932006	14.300998

[1500 rows x 7 columns]

Make 3d Plot

```
matplotlib.pyplot, sklearn.datasets
import pandas as pnd
import numpy as np
import matplotlib.pyplot as plt
import sklearn.datasets as skl_dataset
c_map=plt.cm.get_cmap("YlGnBu")
x_variables,y = skl_dataset.make_friedman1(n_samples=1500,
n_features=6,random_state=0, noise=0.0) # With the make_friedman1 function,
1500 samples are created without any noise
data_frame=pnd.DataFrame(x_variables,columns=['x'+str(i) for i in
range(1,7)])
data_frame['y'] = y
fig = plt.figure(figsize=(7,7)) # Figure size
ax = fig.add_subplot(111, projection='3d')
ax.scatter(data_frame.iloc[:,0], data_frame.iloc[:,1],data_frame.
iloc[:,2],c=data_frame.y, cmap=c_map)# A 3D graph was drawn using the first
three features of the data set created with the help of the function
plt.title('Function: Friedman1') # Title of the graph
plt.show()
```

Output:

You will see the graph shown in Figure 5-7.

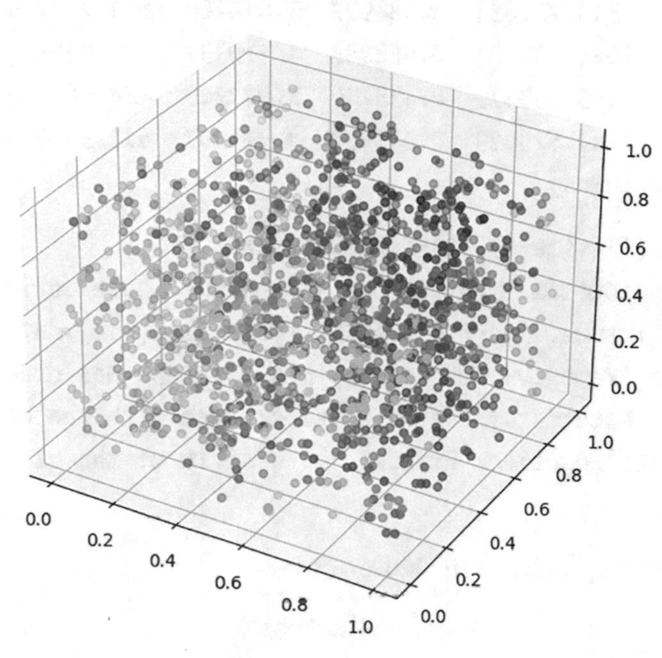

Figure 5-7. *Friedman 1 samples*

make_friedman2() function: make_friedman2 () has 4 input dimensions and one target variable.

```
x_variables,y = skl_dataset.make_friedman2(n_samples=1500,random_state=0,
noise=0.0) # With the make_friedman2 function, 1500 samples are created
without any noise
data_frame2=pnd.DataFrame(x_variables,columns=['x'+str(i) for i in
range(1,5)])
data_frame2['y'] = y
print (data_frame2)
```

Output:

```
          x1           x2        x3        x4            y
0    54.881350  1294.017209  0.602763  6.448832   781.914458
1    42.365480  1180.814530  0.437587  9.917730   518.443136
2    96.366276   752.064577  0.791725  6.288949   603.175873
3    56.804456  1637.744458  0.071036  1.871293   129.465874
4     2.021840  1485.854949  0.778157  9.700121  1156.229758
...        ...          ...       ...       ...          ...
```

```
1495  90.761456  1282.735004  0.528287  7.393068  683.702753
1496  73.083957  1315.726634  0.484977  3.814760  642.268770
1497  20.617459  1626.337223  0.058247  4.370731   96.946340
1498  69.005014  1105.953379  0.433219  2.042329  484.062842
1499  13.952805  1263.523099  0.483697  4.395507  611.320916
[1500 rows x 5 columns]
```

```
fig = plt.figure(figsize=(7,7)) # Figure size
ax = fig.add_subplot(111, projection='3d') # 3D figure definition
ax.scatter(data_frame2.iloc[:,0], data_frame2.iloc[:,1],data_frame2.
iloc[:,2],c=data_frame2.y, cmap=c_map) # A 3D graph was drawn using
the first three features of the data set created with the help of the
functionplt.title('Function: Friedman2') # Title of the graph
plt.show()
```

You will see the graph shown in Figure 5-8.

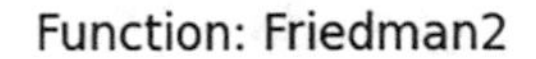

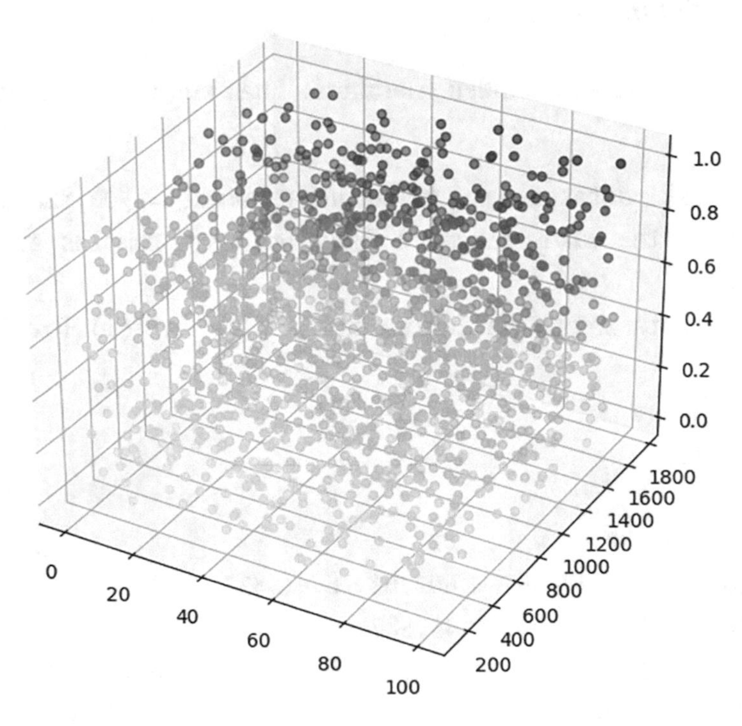

Figure 5-8. *Friedman 2 samples*

make_friedman3() function: make_friedman3 () also has 4 input dimensions and one target variable.

```
x_variables,y = skl_dataset.make_friedman3(n_samples=1500,random_state=0,
noise=0.0) # With the make_friedman3 function, 1500 samples are created
without any noise
data_frame3=pnd.DataFrame(x_variables,columns=['x'+str(i) for i in
range(1,5)])
data_frame3['y'] = y
print (data_frame3)
```

Output:

```
           x1           x2        x3        x4         y
0    54.881350  1294.017209  0.602763  6.448832  1.500550
1    42.365480  1180.814530  0.437587  9.917730  1.488988
2    96.366276   752.064577  0.791725  6.288949  1.410344
3    56.804456  1637.744458  0.071036  1.871293  1.116578
4     2.021840  1485.854949  0.778157  9.700121  1.569048
...        ...          ...       ...       ...       ...
1495 90.761456  1282.735004  0.528287  7.393068  1.437653
1496 73.083957  1315.726634  0.484977  3.814760  1.456759
1497 20.617459  1626.337223  0.058247  4.370731  1.356491
1498 69.005014  1105.953379  0.433219  2.042329  1.427755
1499 13.952805  1263.523099  0.483697  4.395507  1.547970
[1500 rows x 5 columns]
```

Make3d Plot

```
fig = plt.figure(figsize=(7,7)) # Figure size
ax = fig.add_subplot(111, projection='3d') # 3D figure definition
ax.scatter(data_frame3.iloc[:,0], data_frame3.iloc[:,1],data_frame3.
iloc[:,2],c=data_frame3.y, cmap=c_map) # A 3D graph was drawn using the
first three features of the data set created with the help of the function
plt.title('Function: Friedman3') # Title of the graph
plt.show()
```

You will see the graph shown in Figure 5-9.

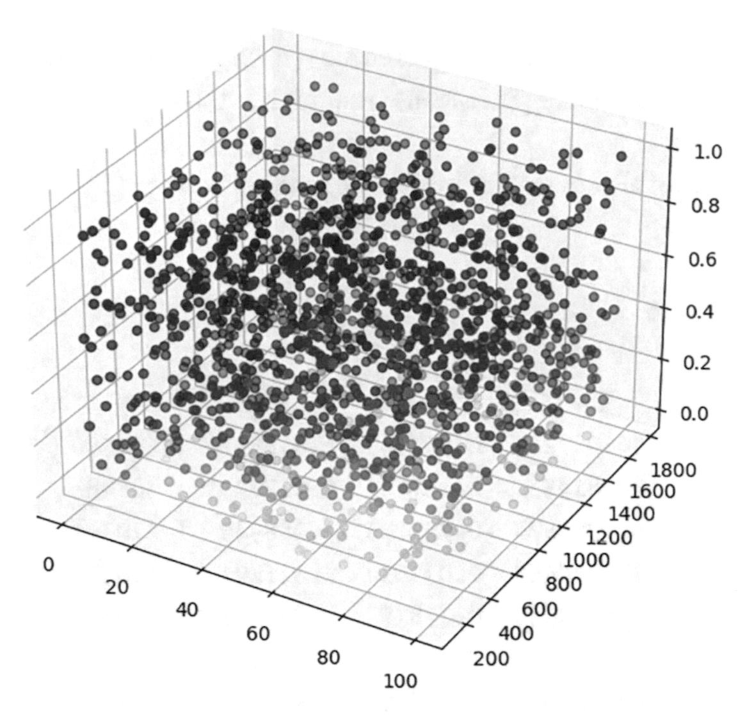

***Figure 5-9.** Friedman 3 samples*

Since these functions, make_friedman, equations are predetermined, no changes can be made in the equation. So, how we are going to generate data when we have our equation? In this case, we use the symbolic regression library. In the symbolic regression library, any equation of the data can be determined, then synthetic data can be produced easily.

The gen_regression_symbolic() function is used to create the synthetic data.

Our symbolic expression is:

$$2x_1 - \frac{x_2^2}{5} + 15 * \cos(x_3) + noise$$

gen_regression_symbolic() is a function that runs under the SymPy library. You can also use direct code, which is shown in `https://github.com/tirthajyoti/Synthetic-data-gen/blob/master/Notebooks/Symbolic%20regression%20classification%20generator.ipynb`.

```
sym_reg = gen_regression_symbolic(m='(2*x1-(x2^2)/5+15*cos(x3))',n_
samples=100,noise=0.001) # Generates 100 samples according to any given
function with small noise value
data_frame=pnd.DataFrame(sym_reg, columns=['x'+str(i) for i in
range(1,4)]+['y'])
print (data_frame)
```

Output:

```
         x1        x2        x3                y
0    4.01501  -7.82078  -3.28809  -19.0413221727414
1   -2.87236   4.67456   2.15993  -18.4489373520397
2   0.583208  -3.63311   1.89078  -6.19090192508419
3   -1.83245   1.34786   3.11348  -19.0215795562634
4    6.07163   4.08164 -0.555099   21.5589808291341
..       ...       ...       ...               ...
95   6.82835  -9.05487  -7.16284   6.82160246144781
96 -0.182998  -2.88295   7.12683   7.94190040809628
97   1.90985   2.00882  -6.91558   15.1138855856051
98  -11.8337  -5.63271  0.291907  -15.6480137151175
99  -4.86859  -6.69322    10.473  -26.1853405176567

[100 rows x 4 columns]
```

```
sym_reg = gen_regression_symbolic(m='(2*x1-(x2^2)/5+15*cos(x3))',n_
samples=100,noise=0.001) # Generates 100 samples according to any given
function with small noise value
data_frame=pnd.DataFrame(sym_reg, columns=['x'+str(i) for i in
range(1,4)]+['y'])
fig, ax = plt.subplots(1,3,figsize=(10,6))
a=131
for i in range(3):
    plt.subplot(a)
    plt.scatter(data_frame[data_frame.columns[i]],data_frame.y,
    s=100,c="red", edgecolor="black")
    plt.title("Symbolic Regression value of x"+str(i+1), size=10)
    plt.grid(True)
    a+=1
plt.show()
```

Output:

You will see the graph shown in Figure 5-10.

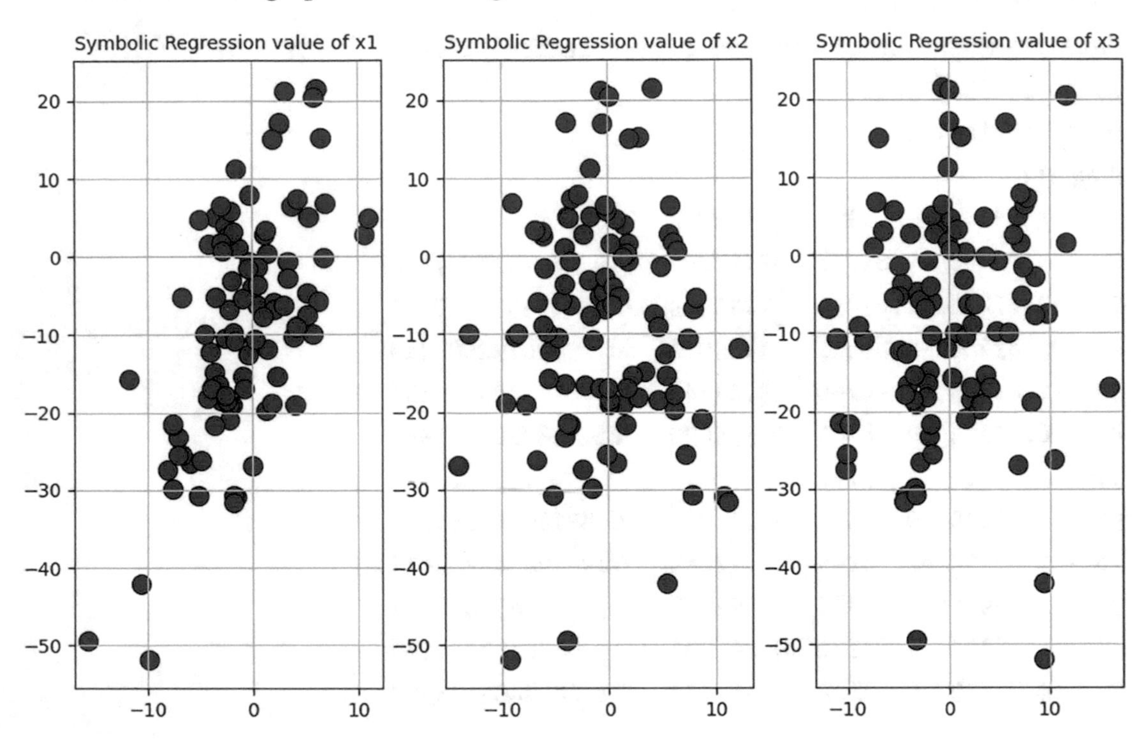

***Figure 5-10.** Symbolic regression samples*

When the noise value increases, new data set plot of the same function is below:

```
sym_reg = gen_regression_symbolic(m='(2*x1-(x2^2)/5+15*cos(x3))',n_
samples=100,noise=100) # Generates 100 samples according to any given
function with noise value=100
data_frame=pnd.DataFrame(sym_reg, columns=['x'+str(i) for i in
range(1,4)]+['y'])
print (data_frame)
```

Output:

	x1	x2	x3	y
0	-3.46277	-5.51495	3.98618	-57.4350979414258
1	5.43253	-4.04728	9.11698	9.37138073808718
2	-6.31532	2.5455	9.62771	-160.316068073531
3	1.82535	1.91883	7.87547	-72.6741141566774

```
4  -4.58646  3.08036 -1.98984  10.6154364431634
..      ...       ...       ...           ...
95  6.22184  -3.52058 -3.46112   27.1446124999385
96 -9.17476  -1.60239  4.05507  -237.823167918562
97  9.55225 -0.942542    5.795   32.2341630424007
98  3.43748  7.74819  4.16235    5.50366250210477
99 -5.28268   2.22154  3.60838   100.090605544604
[100 rows x 4 columns]
sym_reg = gen_regression_symbolic(m='(2*x1-(x2^2)/5+15*cos(x3))',n_
samples=100,noise=100) # Generates 100 samples according to any given
function with noise value=100
data_frame=pnd.DataFrame(sym_reg, columns=['x'+str(i) for i in
range(1,4)]+['y'])
fig, ax = plt.subplots(1,3,figsize=(10,6)) # Generating figures for 3
parameters values consisting of 1 rows 3columns
a=131
for i in range(3):
    plt.subplot(a)
    plt.scatter(data_frame[data_frame.columns[i]],data_frame.y,
    s=100,c="red", edgecolor="black")
    plt.title("Symbolic Regression value of x"+str(i+1), size=10)
    plt.grid(True)
    a+=1

plt.show()
```

Output:

You will see the graph shown in Figure 5-11.

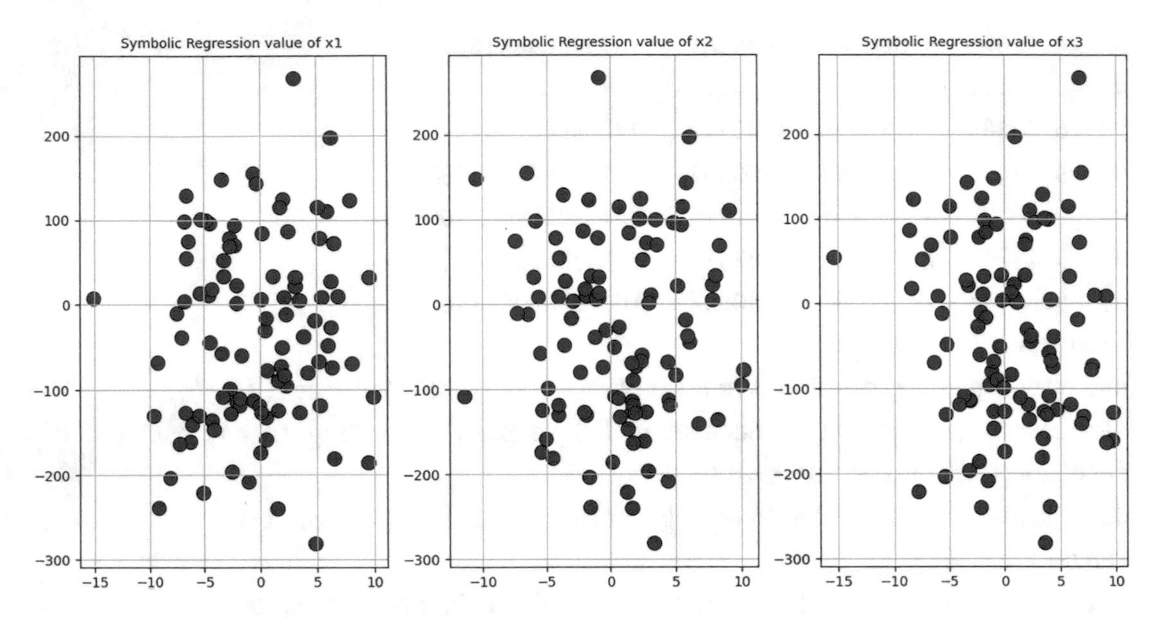

Figure 5-11. *Symbolic regression samples with noise*

Synthetic data generation for Classification and Clustering Problems

There are many ways to generate synthetic data for classification and clustering problems. One popular method is to use a generative model, such as a Gaussian mixture model, to generate data from a known distribution. Another common method is to use a bootstrap approach, which involves sampling data with replacement from a dataset and then fitting a model to the generated data.

Bootstrapping is a powerful method for generating synthetic data that can be used for both classification and clustering problems. The key advantage of using a bootstrap approach is that it allows you to generate data that is representative of the underlying distribution of the data. This is important because it means that the generated data will be useful for training and testing models.

Gaussian mixture models are another popular method for generating synthetic data. These models can generate data that is realistic and that conforms to a known distribution. Gaussian mixture models are especially useful for generating data for clustering problems.

There are many other methods for generating synthetic data, such as using Random Forests or Support Vector Machines. No matter which method you use, the important thing is that the generated data is representative of the real data. This will ensure that the models you train will be effective and will generalize well to new data.

Classification Problems

Scikit-learn is one of the Python libraries to create and analyze data. Classification, regression, clustering, dimension reduction, model selection, and Preprocessing are the main functions in this python library.

Classification problems are going to discuss in this part.

The "make_classification" function is similar to "make_regression" function. The "make_classification" function has many parameters. Samples, features, classification number, number of clusters in each class, flip_y, just a few of them.

Now, data will be produced by changing features of classification and the number of sample parameters. The triple combinations of the generated data will be plotted in 3 dimensions.

```
import pandas as pnd
import numpy as np
import matplotlib.pyplot as plt
from sklearn.datasets import make_classification
from itertools import combinations
from math import ceil
# Many libraries are required for the code
data_class = make_classification(n_samples=150, n_features=4) # 150 samples
with 4 features were generated with the make_classification function.
d_fr = pnd.DataFrame(data_class[0], columns=["x1","x2","x3","x4"]) #
Generate data frame
d_fr['y'] = data_class[1]
print (d_fr.head())
```

Output:

```
         x1        x2        x3        x4  y
0  0.281367  0.347271 -0.006246  0.440627  1
1 -0.411207 -0.855647  0.715458  0.714364  0
2 -0.488541 -0.350263 -0.501890 -1.751094  0
```

```
3  0.594585  0.745571 -0.036975  0.885411  1
4  0.963687  1.234084 -0.112037  1.334840  1
comb_var=list(combinations(d_fr.columns[:-1],3)) # Creating 3-combination
sets from 4-features data set.
print (comb_var)
```

Output:

```
[('x1', 'x2', 'x3'), ('x1', 'x2', 'x4'), ('x1', 'x3', 'x4'), ('x2',
'x3', 'x4')]
lenght_comb = len(comb_var)
fig = plt.figure(figsize=(11,7))
a=221
for ii in range(lenght_comb):
    ax = fig.add_subplot(a+ii, projection='3d') # 3D figure definition
    x1 = comb_var[ii][0]
    x2 = comb_var[ii][1]
    x3 = comb_var[ii][2]
    ax.scatter3D(d_fr[x1],d_fr[x2],d_fr[x3],c=d_fr['y'],edgecolor='b',
    s=100) # 3D Scatter plot
    plt.title('Variables'+str(comb_var[ii]))
    plt.grid(True)
plt.show()
```

Output:

You will see the graph shown in Figure 5-12.

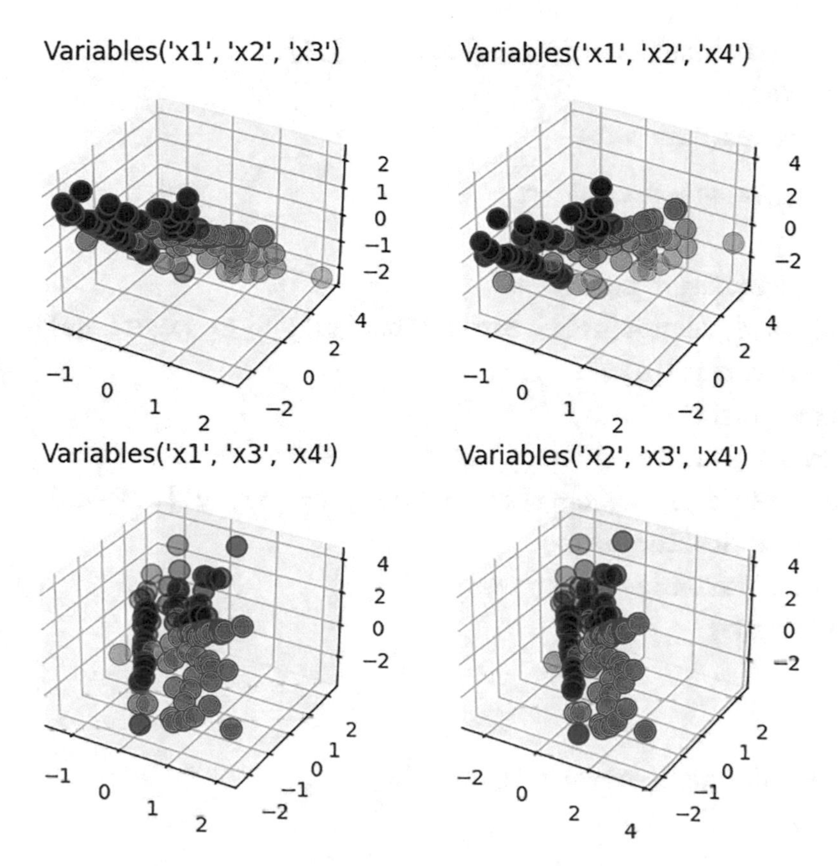

Figure 5-12. Classification function samples

We can easily define the separation degree of classes by using the "class_sep" features of the "make_classification" function.

```
data_class = make_classification(n_samples=200, n_features=4, class_
sep=5.0)
d_fr = pnd.DataFrame(data_class[0], columns=["x1","x2","x3","x4"])
d_fr['y'] = data_class[1]
print (d_fr.head())
Output:
         x1        x2        x3        x4  y
0  4.770371 -3.957778 -4.727696  2.055952  0
1  6.520983 -5.950567 -4.329821  2.218751  0
2 -5.875455  5.061547  5.085108 -2.327549  1
3 -4.863805  6.492111 -4.876516  0.593860  1
4  5.955333 -4.964671 -5.808199  2.540613  0
```

```
comb_var=list(combinations(d_fr.columns[:-1],3))
print (comb_var)
lenght_comb = len(comb_var)

fig = plt.figure(figsize=(11,7))
a=221
for ii in range(lenght_comb):
    ax = fig.add_subplot(a+ii, projection='3d') # 3D figure definition
    x1 = comb_var[ii][0]
    x2 = comb_var[ii][1]
    x3 = comb_var[ii][2]
    ax.scatter3D(d_fr[x1],d_fr[x2],d_fr[x3],c=d_fr['y'],edgecolor='b',
    s=100) # 3D Scatter plot
    plt.title('Variables'+str(comb_var[ii]))
    plt.grid(True)
plt.show()
```

Output:

You will see the graph shown in Figure 5-13.

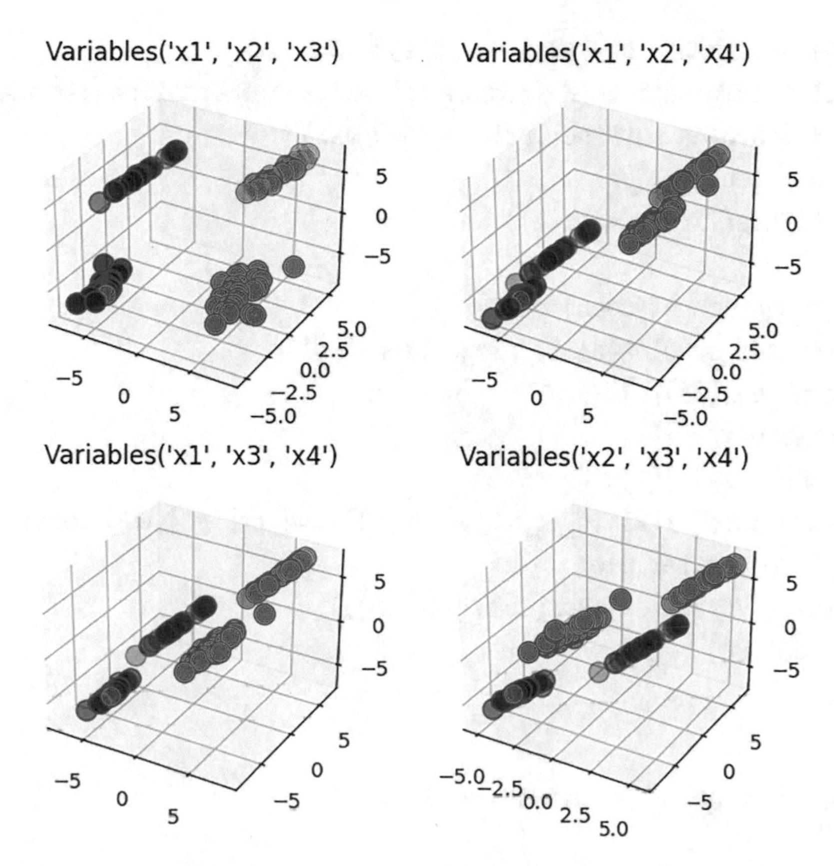

Figure 5-13. *Classification function with 5.0 separation degree*

The default class separation value is 1. By reducing the class_sep value, class separation can be made difficult. Thus, we can control the degree of difficulty of the class separation.

```
data_class = make_classification(n_samples=200, n_features=4, class_
sep=0.01)
d_fr = pnd.DataFrame(data_class[0],columns=["x1","x2","x3","x4"])
d_fr['y'] = data_class[1]
print (d_fr.head())
Output:
          x1        x2        x3        x4  y
0 -1.088551  0.149864 -1.166086  0.635055  1
1  0.582877 -0.523155  1.048317 -0.695005  0
2 -0.637509 -0.384611 -0.230787 -0.006657  0
3  0.246752 -0.653592  0.857389 -0.640534  1
```

```
4  0.809663 -0.228885  0.979717 -0.566453  0
comb_var=list(combinations(d_fr.columns[:-1],3)) # Creating 3-combination
sets from 4-features data set.print (comb_var)
lenght_comb = len(comb_var)
fig = plt.figure(figsize=(11,7))
a=221
for ii in range(lenght_comb):
    ax = fig.add_subplot(a+ii, projection='3d')
    x1 = comb_var[ii][0]
    x2 = comb_var[ii][1]
    x3 = comb_var[ii][2]
    ax.scatter3D(d_fr[x1],d_fr[x2],d_fr[x3],c=d_fr['y'],edgecolor='b',
    s=100) #3D scatter plot
    plt.title('Variables'+str(comb_var[ii]))
    plt.grid(True)
plt.show()
```

Output:

You will see the graph shown in Figure 5-14.

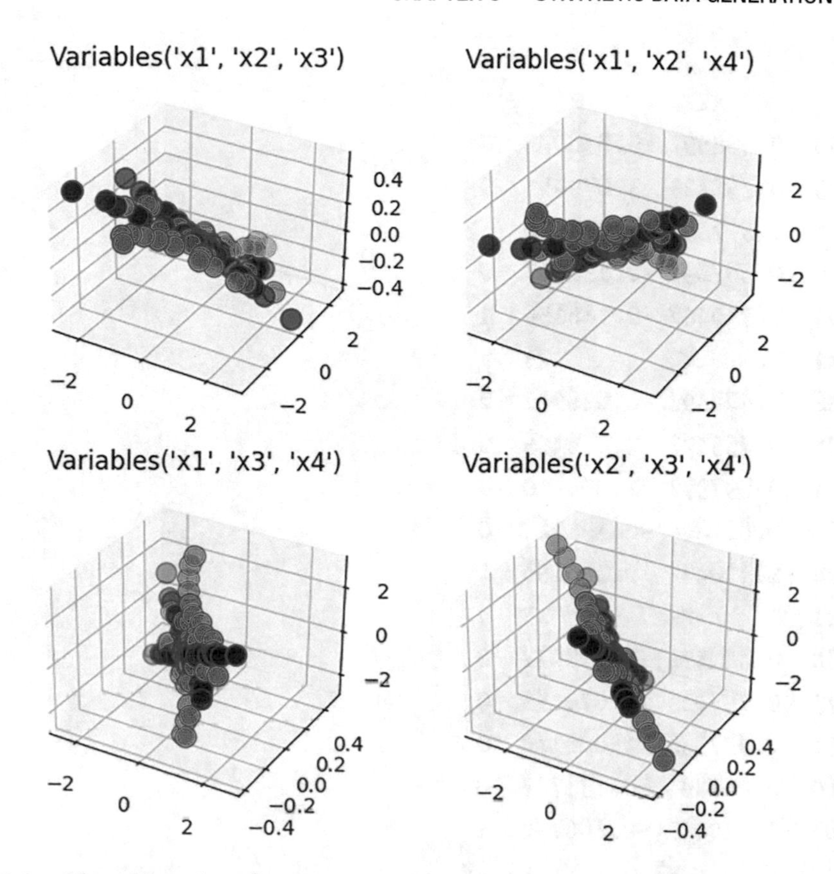

***Figure 5-14.** Classification function with 0.01 separation degree*

Plots are prepared according to different class_separation parameters.

```
c_map=plt.cm.get_cmap("YlGnBu") # Colur of the figure
fig, ax = plt.subplots(3,2,figsize=(11,6))
a=221
sep_par=[0.5,1,5,10] # Different separation parameters
for i in range(4):
    data_class = make_classification(n_samples=100,class_sep=sep_par[i],
    n_features=3,n_informative=1,n_clusters_per_class=1,n_redundant=0,
    random_state=99) # Creates a data set for each separation parameter.
    d_fr = pnd.DataFrame(data_class[0], columns=["x1","x2","x3"])
    d_fr['y'] = data_class[1]
    print (d_fr.head())
```

```
Output:
         x1        x2        x3  y
0 -0.719888  0.488592 -0.838072  0
1 -0.641575  0.755223 -3.079455  0
2 -0.240191  0.997332  1.006110  0
3 -0.542860  0.667894 -0.131717  0
4  0.551964 -1.310447 -0.186156  1
         x1        x2        x3  y
0 -1.219888  0.488592 -0.838072  0
1 -1.141575  0.755223 -3.079455  0
2 -0.740191  0.997332  1.006110  0
3 -1.042860  0.667894 -0.131717  0
4  1.051964 -1.310447 -0.186156  1
         x1        x2        x3  y
0 -5.219888  0.488592 -0.838072  0
1 -5.141575  0.755223 -3.079455  0
2 -4.740191  0.997332  1.006110  0
3 -5.042860  0.667894 -0.131717  0
4  5.051964 -1.310447 -0.186156  1
          x1        x2        x3  y
0 -10.219888  0.488592 -0.838072  0
1 -10.141575  0.755223 -3.079455  0
2  -9.740191  0.997332  1.006110  0
3 -10.042860  0.667894 -0.131717  0
4  10.051964 -1.310447 -0.186156  1
    plt.subplot(a)
    plt.scatter(d_fr["x1"],d_fr['x2'],c=d_fr['y'], s=100) # Scatter plot
    for each separation parameter.
    plt.title('Class Separation='+ str(sep_par[i]), size=10)
    plt.grid(True)
    a+=1
plt.show()
```

Output:

You will see the graph shown in Figure 5-15.

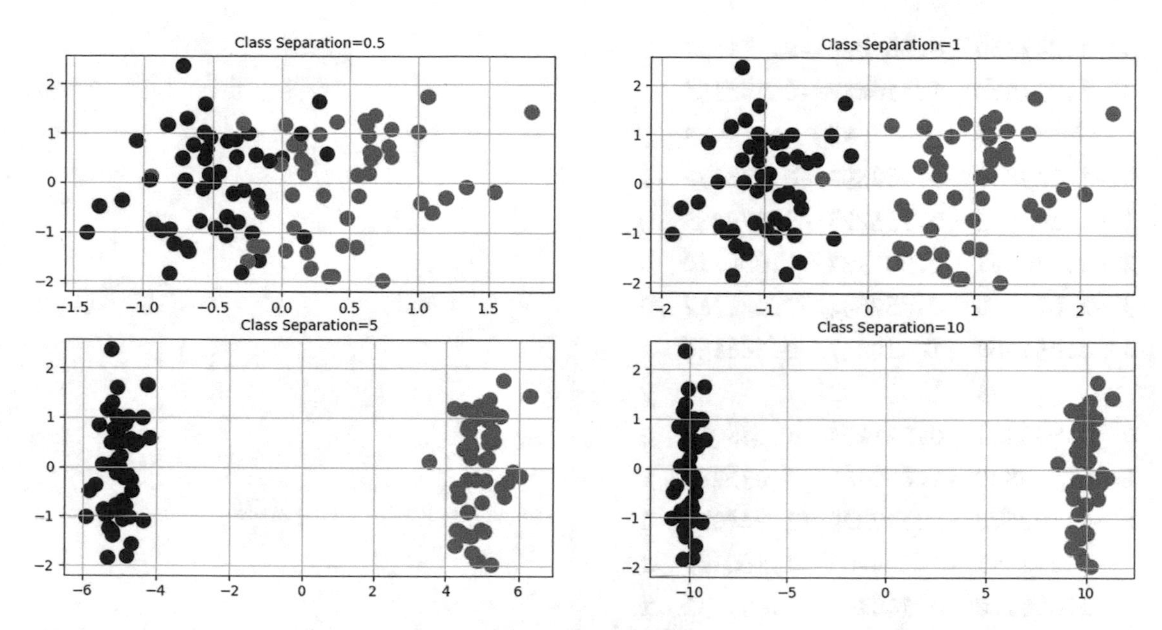

***Figure 5-15.** Classification function with a different separation degree*

Additionally, we may want to control the noise in the data with flip_y features. The default value of it is 0.01, by increasing this value, we can make noisier data.

```
c_map=plt.cm.get_cmap("YlGnBu") # Color of the figure
fig, ax = plt.subplots(3,2,figsize=(11,6)) # Figure definition with 3 rows
and 2 columns for 6 different noise parametersa=231
noise_data=[0.01,0.1,0.3,0.5,0.75,1]
for i in range(6):
    data_class = make_classification(n_samples=100,flip_y=noise_data[i],
    n_features=3,n_informative=1,n_clusters_per_class=1,n_redundant=0,
    random_state=99) # 100 samples with 3 features were generated with the
    make_classification function.
    d_fr = pnd.DataFrame(data_class[0], columns=["x1","x2","x3"])
    d_fr['y'] = data_class[1]
    print (d_fr.head())
Output:
         x1        x2        x3  y
0 -1.219888  0.488592 -0.838072  0
1 -1.141575  0.755223 -3.079455  0
2 -0.740191  0.997332  1.006110  0
```

```
3 -1.042860  0.667894 -0.131717  0
4  1.051964 -1.310447 -0.186156  1
         x1        x2        x3  y
0 -1.219888  0.488592 -0.838072  0
1 -1.141575  0.755223 -3.079455  0
2 -0.740191  0.997332  1.006110  0
3 -1.042860  0.667894 -0.131717  0
4  1.051964 -1.310447 -0.186156  1
         x1        x2        x3  y
0  1.186402  0.221004  1.188059  0
1  0.529890 -1.200443  0.036284  1
2 -0.402620 -1.277329 -1.244909  1
3 -0.212651 -1.324401  1.160567  0
4  1.006110 -0.740191  0.997332  1
         x1        x2        x3  y
0 -1.739733  0.718727  0.558044  1
1 -1.849785 -1.309492  0.739926  1
2  0.362326  0.493240  1.443892  1
3 -0.486586 -1.811083 -2.157206  0
4 -1.282523  0.299280  1.804544  1
         x1        x2        x3  y
0 -0.962169 -1.451818  0.044653  1
1 -0.733295 -0.827071  0.512437  0
2 -0.661349  1.568567  1.743235  0
3 -0.444609 -0.980022 -0.923461  0
4 -0.210006 -1.425160 -0.865028  1
         x1        x2        x3  y
0 -1.277345 -1.006062  0.953027  1
1  1.165405  2.233999  0.529370  0
2  1.169992  0.013460  1.126487  1
3 -1.901419  1.020386  0.852078  0
4  1.435830  0.801733  2.311340  1
    plt.subplot(a)
    plt.scatter(d_fr["x1"],d_fr['x2'],c=d_fr['y'], s=100)
    plt.title('Noise='+ str(noise_data[i]), size=10)
```

```
    plt.grid(True)
    a+=1
plt.show() #
```

You will see the graph shown in Figure 5-16.

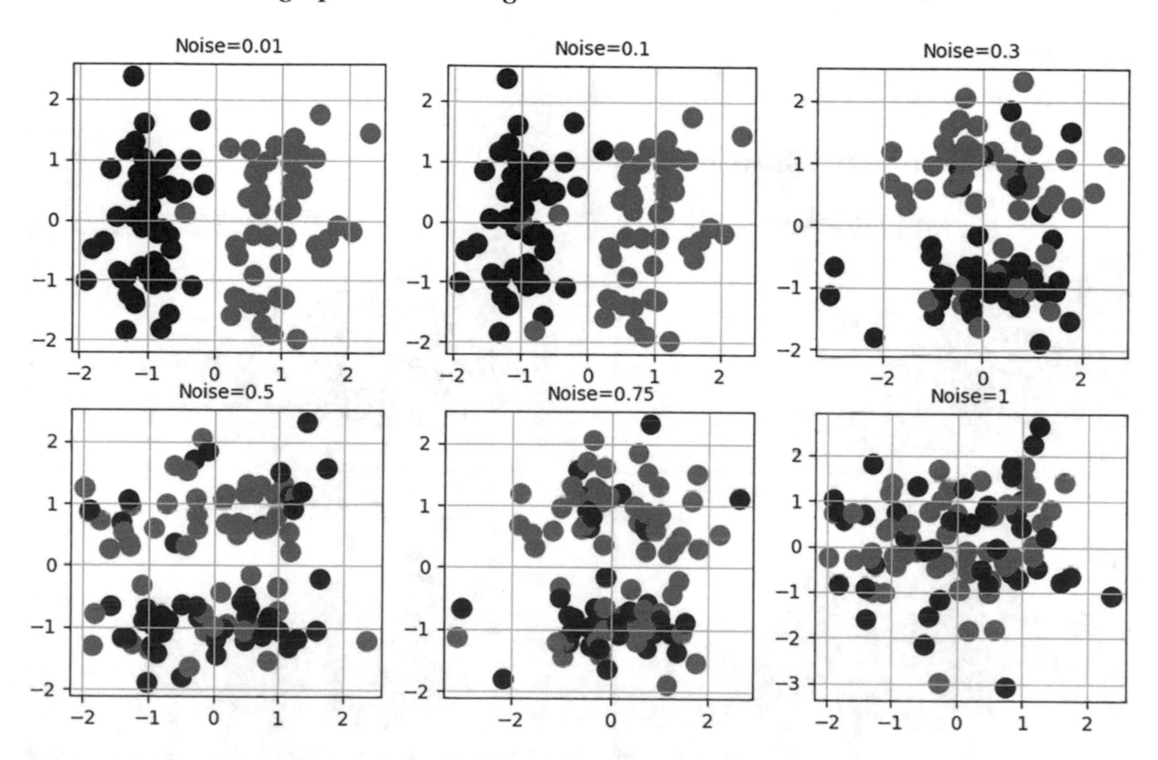

Figure 5-16. *Classification function with different noise*

Multi-label classification is a modeling technique developed to predict the label of a class that has zero or more than one common label.

```
from sklearn.datasets import make_multilabel_classification
c_map=plt.cm.get_cmap("YlGnBu")
fig, ax = plt.subplots(2,2)
a=221
number_labels=[2,5, 7, 10]
for ii in range(4):
    plt.subplot(a)
    x_class, y_class= make_multilabel_classification(n_samples=500,
    n_features=4,random_state=99, n_classes=3,n_labels=number_labels[ii])
    # 500 samples with 4 features were generated based on the labels number
```

```
    new_y=np.sum(y_class*[4,2,1], axis=1)
    plt.scatter(x_class[:,2],x_class[:,3],c=new_y, s=100, cmap=c_map) # The
    scatter plot created using the data in the 3rd and 4th features
    plt.title('Number of Labels='+str(number_labels[ii]))
    a+=1
plt.show()
```

Output:

You will see the graph shown in Figure 5-17.

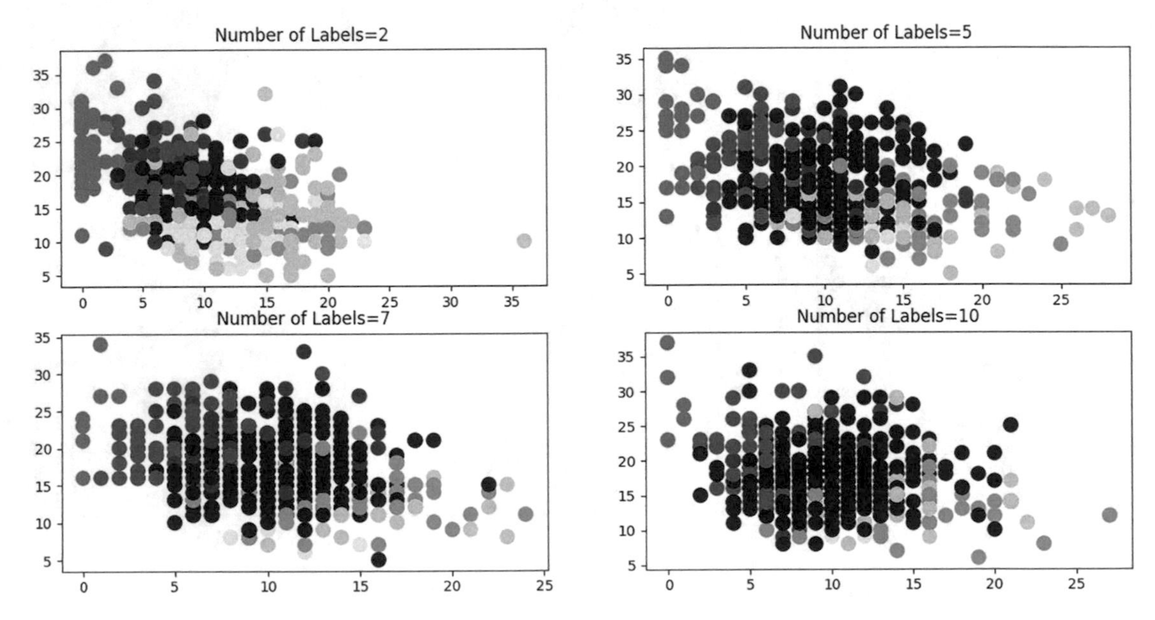

Figure 5-17. *Multi-label classification samples*

Clustering Problems

There are a few different types of clustering problems that can occur when synthetic data is generated. One type of problem is when the data is not evenly distributed among the clusters. This can happen if the data is generated randomly, without considering the underlying structure of the data. Another type of problem is when the clusters are too close together and overlap with each other. This can happen if the data is generated using a small number of clusters or if the clusters are not well-defined. Finally, the third type of problem can occur when the clusters are too far apart and do not share any data points. This can happen if the data is generated using a large number of clusters, or if the clusters are not well-defined.

There are many different types of clustering problems that can occur when synthetic data is generated. Some of the most common problems includes:

- **Overlapping clusters** - This occurs when two or more clusters have generated clusters overlap with each other. This can cause problems when trying to analyze the data as it can be difficult to determine which data points belong to which cluster.
- **Inconsistent cluster boundaries** -This occurs when the boundaries between clusters are not consistent. Thsi can agin make it difficult to analyze the data as it can be hard to determine which data points belong to which cluster.
- **Uneven cluster sizes** -This occurs when the sizes of the clusters are not evenly distributed. This can make it difficult to compare the results of different clusters.
- **Non-uniform cluster shapes** -This occurs when the shapes of the clusters are not uniform. This can make it difficult to compare the results of different clusters.

Data is generated by using the sklearn.datasets library to create data components similar to blob. make_blobs() function generates the Gaussian blobs for clustering.

Data set is plotted with different centers number; the default center number is 3.

```
import sklearn.datasets as skl_dt
c_map=plt.cm.get_cmap("YlGnBu")
fig, ax = plt.subplots(2,2)
a=221
for ii in range(3,7):
    plt.subplot(a)
    x_data, y = skl_dt.make_blobs(n_samples=500,centers=ii,random_state=99)
    # 500 samples were generated based on the centers number

    plt.scatter(x_data[:,0],x_data[:,1],c=y, s=50, cmap=c_map)
       plt.title('Number of Centers='+str(ii))
    a+=1
plt.show()
```

Output:

You will see the graph shown in Figure 5-18.

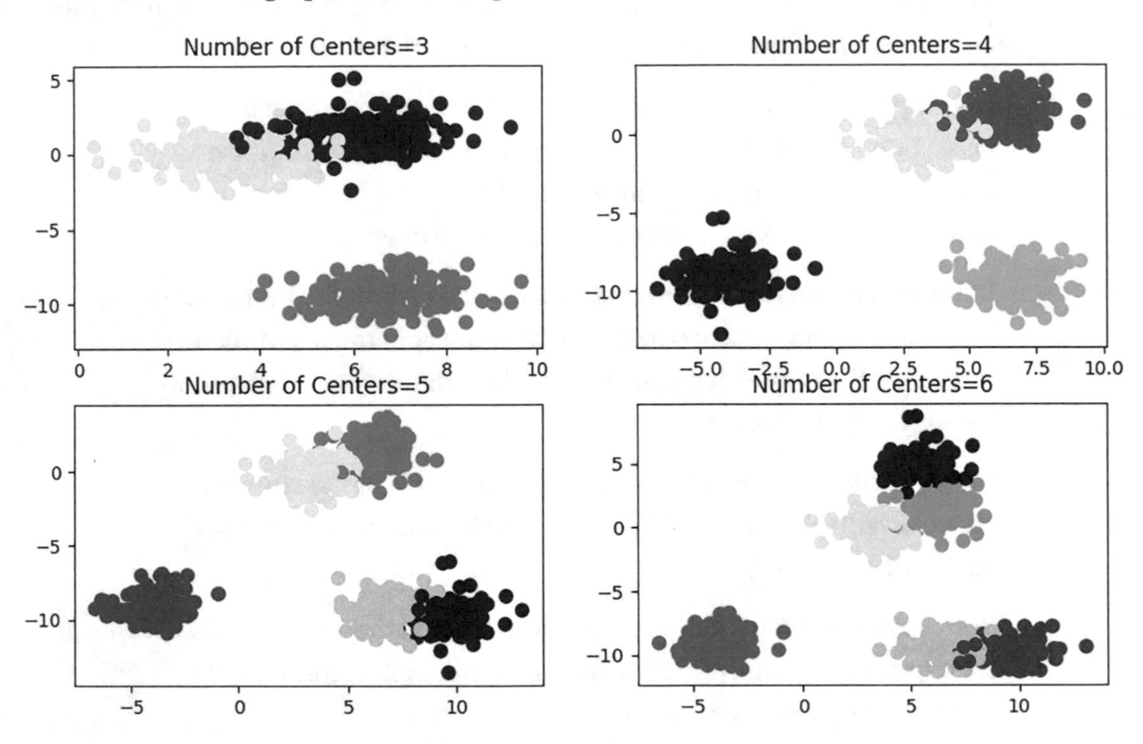

***Figure 5-18.** Gaussian blobs for clustering with different number of centers*

Using 3 centers, we can generate data with 4 features.

```
from itertools import combinations
from math import ceil
c_map=plt.cm.get_cmap("YlGnBu")
fig, ax = plt.subplots(3,2)
data_clust = skl_dt.make_blobs(n_samples=500, n_features=4, centers=3) #
500 samples with 4 features were generated based on the 3 centers.

d_fr = pnd.DataFrame(data_clust[0], columns=["x1","x2","x3","x4"]) # Having
4 features
d_fr['y'] = data_clust[1]
print (d_fr.head())
```

Output:

```
         x1         x2        x3          x4   y
0 -0.330098 -3.499526 -3.932461    1.291953  0
1 -1.096365 -3.965669  2.006255   -6.935994  2
2 -0.888992 -4.453931  2.988597  -10.777764  1
3  0.145460 -3.183180 -3.375980    1.226803  0
4 -1.046599 -4.729893  2.528782   -5.336201  2
```

```
comb_var=list(combinations(d_fr.columns[:-1],2)) )) # Creating
2-combination sets from 4-features data set.
print (comb_var)
[('x1', 'x2'), ('x1', 'x3'), ('x1', 'x4'), ('x2', 'x3'), ('x2', 'x4'),
('x3',x4')]
lenght_comb = len(comb_var)
a=321
for i in range(lenght_comb):

    plt.subplot(a)
    x1 = comb_var[i][0]
    x2 = comb_var[i][1]
    plt.scatter(d_fr[x1],d_fr[x2],c=d_fr['y'],edgecolor='b', s=150)
    plt.xlabel(comb_var[i][0])
    plt.ylabel(comb_var[i][1])
    plt.grid(True)
    a+=1
plt.show()
```

Output:

You will see the graph shown in Figure 5-19.

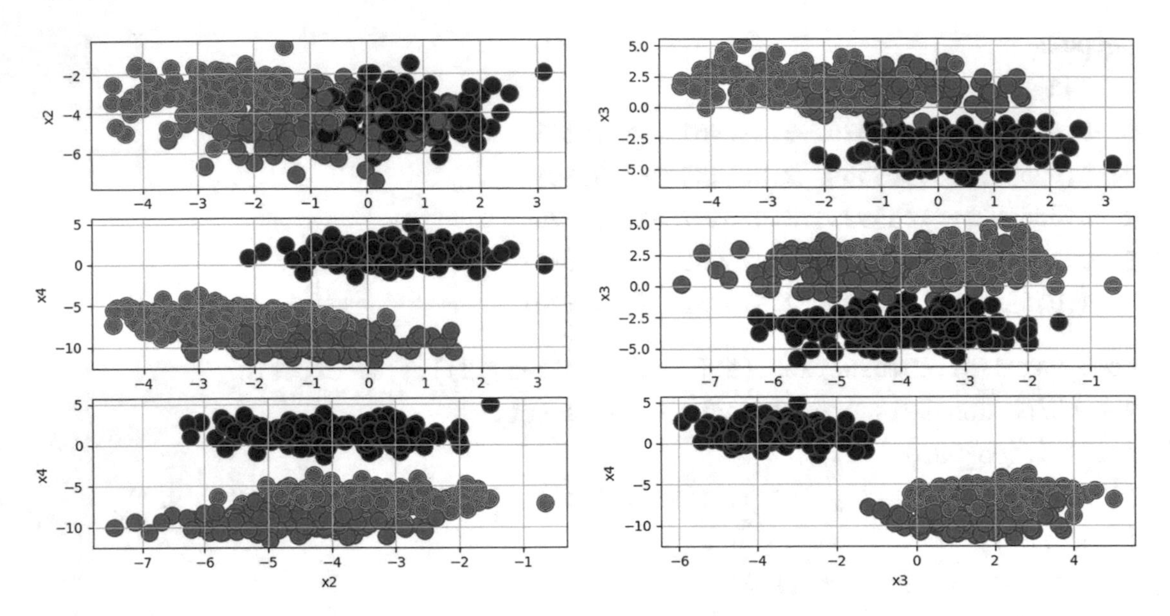

***Figure 5-19.** Gaussian blobs for clustering with different combination of features*

By using the cluster_std feature, we can easily separate our cluster. The default value is 1.

```
import pandas as pnd
import numpy as np
import matplotlib.pyplot as plt
from sklearn.datasets import make_blobs
from itertools import combinations
from math import ceil
c_map=plt.cm.get_cmap("YlGnBu")
fig, ax = plt.subplots(2,2)
cluster_st=[0.3,1,5,10]
a=221
for i in range(4):
    data_clust = make_blobs(n_samples=500, n_features=4, centers=3,
    cluster_std=cluster_st[i]) # 500 samples with 4 features were generated
    based on the 3 centers and specified cluster_st.
    d_fr = pnd.DataFrame(data_clust[0], columns=["x1","x2","x3","x4"])
    d_fr['y'] = data_clust[1]
    plt.subplot(a)
```

```
    plt.scatter(d_fr["x1"],d_fr["x2"],c=d_fr['y'],edgecolor='b', s=150) #
    Scatter plot
    plt.xlabel("x1")
    plt.ylabel("x2")
    plt.title('Cluster sdt='+str(cluster_st[i]))
    plt.grid(True)
    a+=1
plt.show()
```

Output:

You will see the graph shown in Figure 5-20.

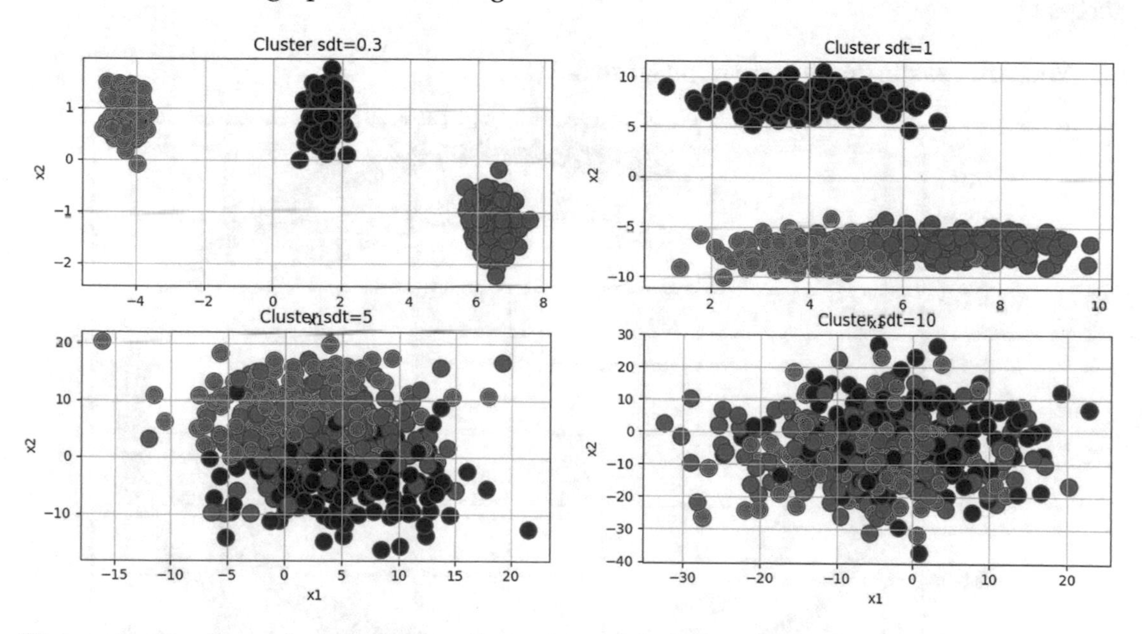

Figure 5-20. *Gaussian blobs for clustering with different cluster separations*

In this sklearn.datasets library, it is also possible to generate data that has a certain shape. Circle and halfcircle shape are mentioned in below.

The make_circles() function generates two circle classes data for the classification problem of machine learning. Number of samples, and the noise level of the data are two important parameters of the function.

```
import sklearn.datasets as skl_dt
c_map=plt.cm.get_cmap("YlGnBu")
data_circle=skl_dt.make_circles(n_samples=200) # Generating data to form a
circle with 200 samples and without any noise
```

```
df_circle = pnd.DataFrame(data_circle[0],columns=["x1", "x2"]) # Data set
has 2 features
df_circle['y'] =data_circle[1]
plt.figure()
plt.scatter(df_circle['x1'],df_circle['x2'],c=df_circle['y'],s=100,edgecolo
rs='k') # Scatter plot for each feature
plt.xlabel('x1')
plt.ylabel('x2')
plt.grid(True)
plt.show()
```

Output:

You will see the graph shown in Figure 5-21.

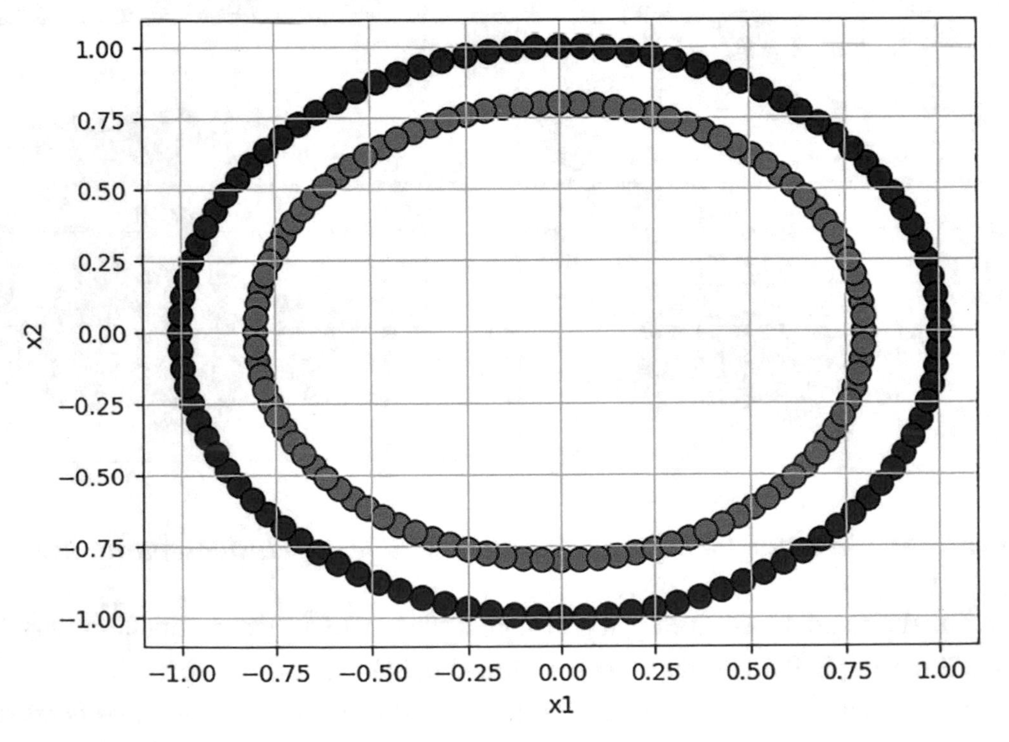

Figure 5-21. *Circle without noise*

With different noise level:

```
import sklearn.datasets as skl_dt
fig, ax = plt.subplots(2,2)
```

```
a=221
for noise_ in [0,0.05,0.1,0.5]: #noise parameter
    plt.subplot(a)
    data_circle=skl_dt.make_circles(n_samples=200, noise=noise_) #
    Generates the data with the noise parameter.
    df_circle = pnd.DataFrame(data_circle[0],columns=["x1", "x2"])
    df_circle['y'] =data_circle[1]
    plt.scatter(df_circle['x1'],df_circle['x2'],c=df_circle['y'],s=100,
    edgecolors='k')
    plt.title('Noise value= '+str(noise_))
    a+=1
plt.show()
```

Output:

You will see the graph shown in Figure 5-22.

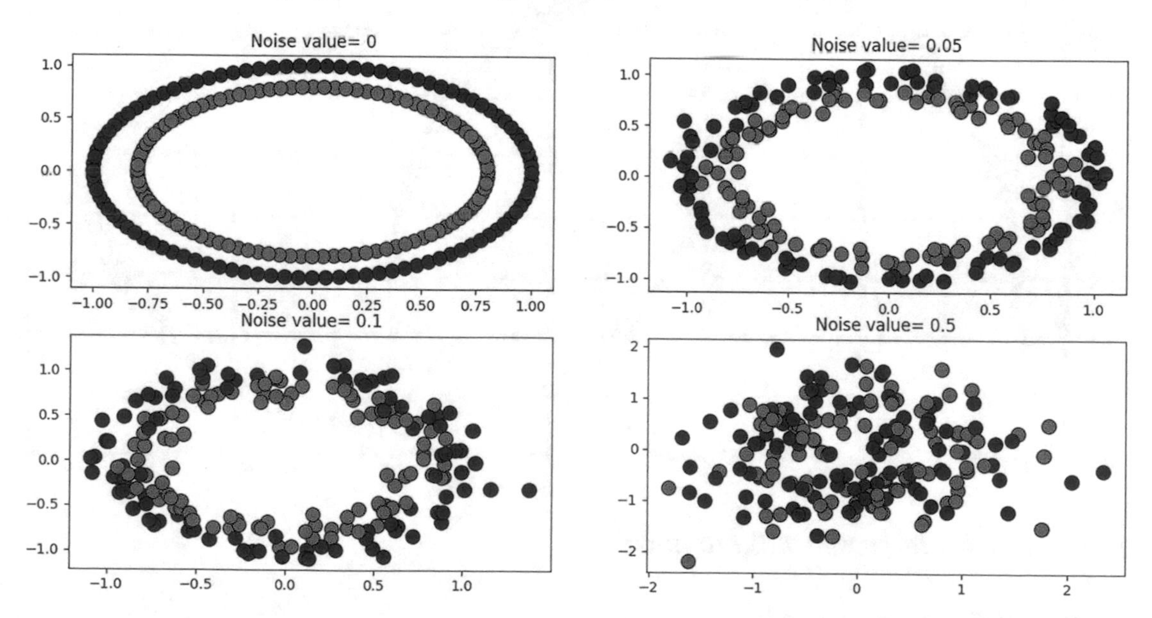

Figure 5-22. *Circle with different noise*

Also, a half-circle can make with the make_moons() function.

```
import sklearn.datasets as skl_dt
data_moon=skl_dt.make_moons(n_samples=200) # Generating data to form a half
circle with 200 samples and without any noise
```

```
df_moon = pnd.DataFrame(data_moon[0],columns=["x1", "x2"]) # Data set has 2
features
df_moon['y'] =data_moon[1]
plt.figure()
plt.scatter(df_moon['x1'],df_moon['x2'],c=df_
moon['y'],s=100,edgecolors='k') #scatter plor for each feature
plt.xlabel('x1')
plt.ylabel('x2')
plt.grid(True)
plt.show()
```

Output:

You will see the graph shown in Figure 5-23.

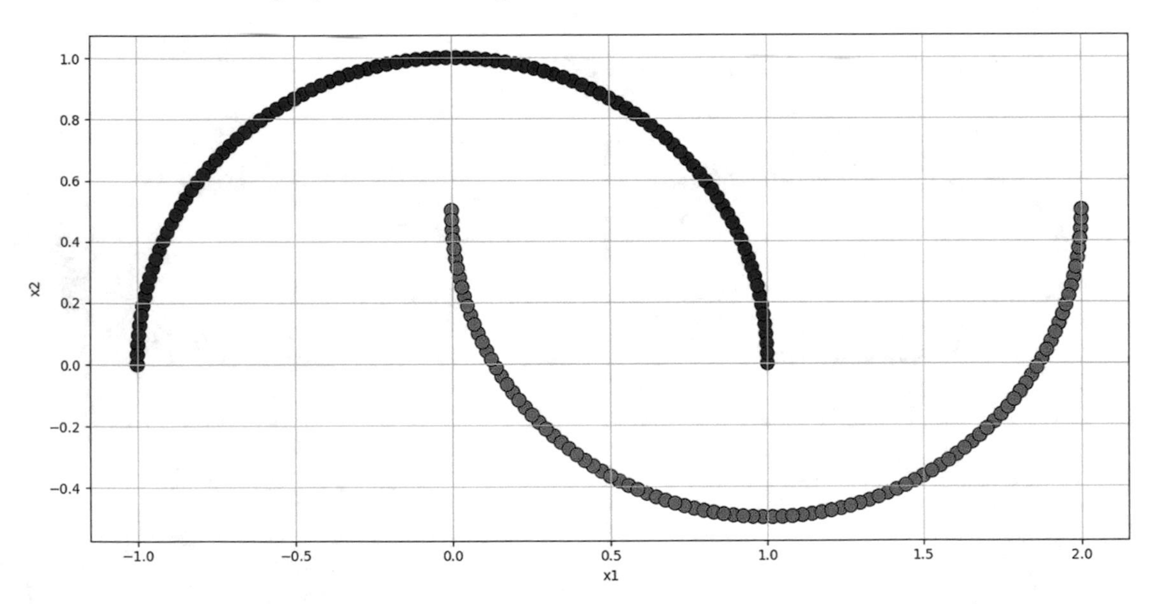

***Figure 5-23.** Half circle without noise*

With different noise levels:

```
import sklearn.datasets as skl_dt
fig, ax = plt.subplots(2,2)
a=221
for noise_ in [0,0.05,0.1,0.5]: # Noise parameter
    plt.subplot(a)
```

```
    data_moon=skl_dt.make_moons(n_samples=200, noise=noise_) # Generates
    the half circle data with the noise parameter.
    df_moon = pnd.DataFrame(data_moon[0],columns=["x1", "x2"])
    df_moon ['y'] =data_moon [1]
    plt.scatter(df_moon['x1'],df_moon ['x2'],c=df_moon ['y'],s=100,
    edgecolors='k')
    plt.title('Noise value= '+str(noise_))
    a+=1
plt.show()
```

Output:

You will see the graph shown in Figure 5-24.

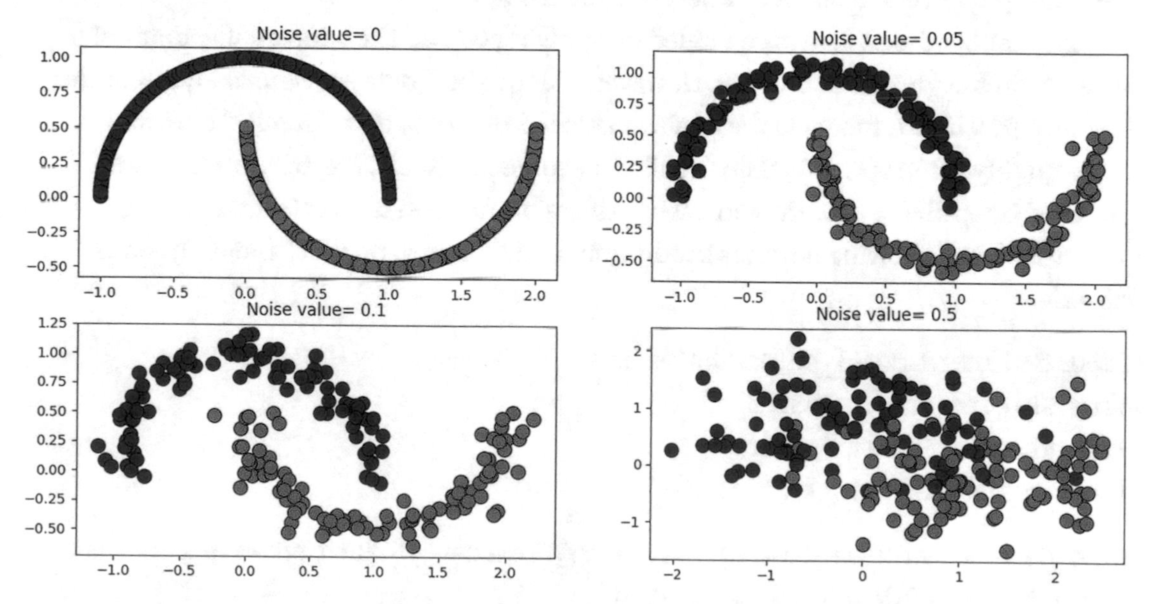

Figure 5-24. *Half circle with different noise*

Generation Tabular Synthetic Data by Applying GANs

GANs can be used to generate synthetic data for any type of data, including images, text, and tabular data. When generating synthetic data, GANs can be used to control for various factors, such as class imbalance, to make the synthetic data more realistic.

One advantage of using GANs to generate synthetic data is that it can be used to create data that is not available. For example, if there is a dataset of images of cats and dogs, but there are no images of rabbits, a GAN could be used to generate synthetic images of rabbits. Another advantage of using GANs to generate synthetic data is that it can be used to protect the privacy of real data. For example, if a dataset contains sensitive information, such as medical records, a GAN could be used to generate synthetic data that is similar to the real data but does not contain any sensitive information.

There are also some disadvantages of using GANs to generate synthetic data. There are a few disadvantages of using GANs to generate synthetic data. First, GANs can be difficult to train. Second, GANs can be slow to generate data. Third, GANs can sometimes generate data that does not look realistic. Despite these disadvantages, GANs are a powerful tool for generating synthetic data. GANs are a promising approach for generating synthetic data for a variety of applications.

Tabular Synthetic data is generated by applying GANs. The wine quality data set is taken from Kaggle public data set. However, the "quality" data set were analyzed under two groups 0 and 1, in which the quality value is 3 or 4 or 5, then "quality" becomes 0, if the quality value is 6, or 7, then "quality" becomes 1. Firstly, the accuracy rate was checked by applying a random forest classification method to a real data set.

Added the following libraries in addition to the Numpy, pandas, and matplotlib libraries

```
import sklearn.model_selection as sms
from sklearn import ensemble
from sklearn import metrics
from tensorflow import keras
```

```
wine_data = pnd.read_csv('C:/..../WineQTNew.csv')# Path shows place that
where the WineQTNew file is located
x_var_name=['fixed_acidity', 'volatile_acidity', 'citric_acid',
'residua_ sugar', 'chlorides', 'free_sulfur_dioxide', 'total_sulfur_dioxide',
'density', 'pH', 'sulphates', 'alcohol']
y_var_name=['quality']
x_wine =wine_data[x_var_name]
y_wine =wine_data[y_var_name]
x_r_trn, x_r_tst, y_r_trn, y_r_tst=sms.train_test_split(x_wine,
y_wine,random_state=40)
random_forest=ensemble.RandomForestClassifier(n_estimators=200)
```

```
random_forest.fit(x_r_trn, y_r_trn.values.ravel())
y_r_prd=random_forest.predict(x_r_tst)

print ("Acurracy: ", metrics.accuracy_score(y_r_tst, y_r_prd))
print ("Classification Result : ",metrics.classification_report(y_r_tst,
y_r_prd))
```

Output:

```
Accuracy:  0.7692307692307693
Classification Result :                    precision    recall  f1-score
support

           0       0.71      0.76      0.73       119
           1       0.82      0.78      0.80       167

    accuracy                           0.77       286
   macro avg       0.76      0.77      0.76       286
weighted avg       0.77      0.77      0.77       286
```

The accuracy of the wine quality model, based on Random Forest Classification, is 0.76. This result will be compared with the model that we trained from generated fake data in the following.

Synthetic data Generation

The training process of a GAN involves the generator and discriminator networks competing against each other to generate realistic data samples. The generator network tries to generate data samples that are realistic enough to fool the discriminator network, while the discriminator network tries to distinguish between real and generated data samples. The competition between the two networks drives the training process and ultimately results in the generator netwotk being able to generate realistic data samples.

There are many applications for GANs, such as image generation, video generation, and text generation. GANs have also been used for more practical applications such as generating realistic images of faces, generating synthetic medical images, and creating new products.

GANs had two main parts, one is the generation the other one is discrimination. The following functions are used for GANs fake data generation [1].

```
def hidden_genset(hidden_size, number_s):
    ent_x = npy.random.randn(hidden_size * number_s)
    ent_x = ent_x .reshape(number_s, hidden_size)
    return ent_x
def fake_generate(genset, hidden_size, number_s):
    ent_x = hidden_genset(hidden_size, number_s)
    fake_x = genset.predict(ent_x)
    fake_y = npy.zeros((number_s, 1))
    return fake_x, fake_y
def real_generate(s_number):
    real_x = wine_data.sample(s_number)
    real_y = npy.ones((s_number, 1))
    return real_x, real_y
def gan_genset(hidden_size, n_outputs=12):
    gan_mdl = keras.models.Sequential()
    gan_mdl.add(keras.layers.Dense(16, activation='selu',
    kernel_initializer='random_normal', input_dim=hidden_size))
    gan_mdl.add(keras.layers.Dense(32, activation='selu'))
    gan_mdl.add(keras.layers.Dense(n_outputs, activation='softmax'))
    return gan_mdl
genset1 = gan_genset(13, 12)
genset1.summary()
```

Output:

```
Model: "sequential"
_________________________________________________________________
 Layer (type)                Output Shape              Param #
=================================================================
 dense (Dense)               (None, 16)                224

 dense_1 (Dense)             (None, 32)                544


 dense_2 (Dense)             (None, 12)                396


=================================================================
```

```
Total params: 1,164
Trainable params: 1,164
Non-trainable params: 0
```

The definition of the discrimination functionis below.

```
def gan_sorter(in_number=12):
    gan_mdl = keras.models.Sequential()
    gan_mdl.add(keras.layers.Dense(30, activation='selu',
    kernel_initializer='random_normal', input_dim=in_number))
    gan_mdl.add(keras.layers.Dense(60, activation='selu'))
    gan_mdl.add(keras.layers.Dense(1, activation='sigmoid'))
    gan_mdl.compile(loss='binary_crossentropy', optimizer='adam',
    metrics=['accuracy'])

    return gan_mdl
sorter1 = gan_sorter(12)
sorter1.summary()
```

```
Model: "sequential_1"

_________________________________________________________________
 Layer (type)                Output Shape              Param #
=================================================================
 dense_3 (Dense)             (None, 30)                390

 dense_4 (Dense)             (None, 60)                1860


 dense_5 (Dense)             (None, 1)                 61

=================================================================
Total params: 2,311
Trainable params: 2,311
Non-trainable params: 0
```

After running the generator and discriminator functions, the GANs function consisting of these functions can be created. The generating GAN model is recalculated each time according to the batch function.

```
def gan_model(genset, sorter):
```

```
    sorter.trainable = False
    gan_mdl = keras.models.Sequential()
    gan_mdl.add(genset)
    gan_mdl.add(sorter)
    gan_mdl.compile(loss='binary_crossentropy', optimizer='adam')
    return gan_mdl

def plot_history(sorter_graph, genset_graph):
    plt.plot(sorter_graph, label='Sorter')
    plt.plot(genset_graph, label='Genset')
    plt.show()
    plt.close()

def data_train(genset_mdl,sorter_model, model_gan, hidden_dim, n_
epochs=1000, n_batch=140, n_eval=250):
    b_size = int(n_batch / 2)
    sorter_hist= []
    genset_hist= []

    for i in range(n_epochs):
        x_r, y_r = real_generate(b_size)
        x_f, y_f = fake_generate(genset_mdl, hidden_dim,b_size)
        r_loss_d, acc_real_d= sorter_model.train_on_batch(x_r, y_r)
        f_loss_d, acc_fake_d= sorter_model.train_on_batch(x_f, y_f)
        loss_value_d = 0.5 * npy.add(r_loss_d,f_loss_d)
        x_g_values = hidden_genset(hidden_dim, n_batch)
        y_g_values = npy.ones((n_batch, 1))
        fake_g_loss = model_gan.train_on_batch(x_g_values, y_g_values)
        print('>%d_value, d1_value=%.4f, d2_value=%.4f d_value=%.4f
        g_value=%.4f' % (i+1, r_loss_d, f_loss_d, loss_value_d, fake_g_
        loss))          sorter_hist.append(loss_value_d)
        genset_hist.append(fake_g_loss )
    plot_history(sorter_hist, genset_hist)
    genset_mdl.save('generated model of trained data.h5') # This is where
    we save the data that we use later
hidden_dim = 13
```

```
sorter = gan_sorter() # gan_sorter function is running
genset = gan_genset(hidden_dim) # gan_genset function is running
new_g_model = gan_model(genset, sorter) # gan_model function is running
data_train(genset, sorter, new_g_model, hidden_dim) # Data train function
is running
```

Output:

```
....
>995_value, d1_value=0.0000, d2_value=0.0001 d_value=0.0000 g_value=9.3510
>996_value, d1_value=0.0000, d2_value=0.0001 d_value=0.0000 g_value=9.3403
>997_value, d1_value=0.0000, d2_value=0.0001 d_value=0.0000 g_value=9.3636
>998_value, d1_value=0.0000, d2_value=0.0001 d_value=0.0000 g_value=9.3703
>999_value, d1_value=0.0000, d2_value=0.0001 d_value=0.0000 g_value=9.3455
>1000_value, d1_value=0.0000, d2_value=0.0001 d_value=0.0000 g_value=9.4020
```

The trained GAN model is saved for later use.

The graph of genset and sorter functions is given in Figure 5-25.

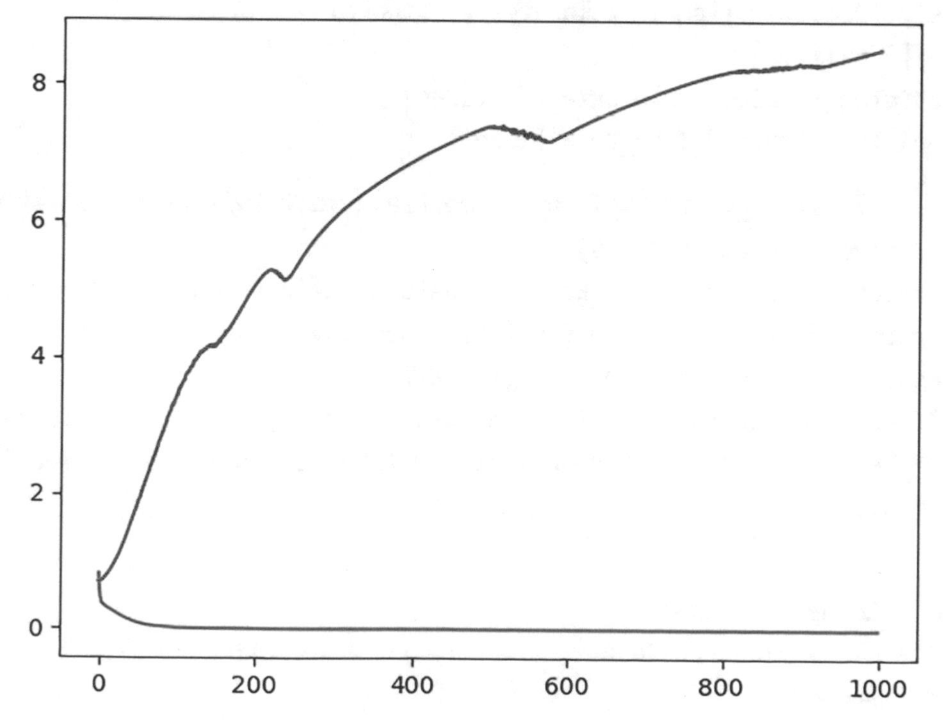

***Figure 5-25.** Sorter and Genset graph*

```
Genset and sorter functions plot.

gan_model_trained =keras.models.load_model('C:/........./generated model of
trained data.h5') # We read our data once file from wherever we saved it.
We have to have specific path.
hidden_dots = hidden_genset(13, 800)
x_predict = gan_model_trained.predict(hidden_dots)
trained_fake_data= pnd.DataFrame(data=x_predict,  columns=['fixed_acidity',
        'volatile_acidity', 'citric_acid', 'residua_ sugar','chlorides',
        'free_sulfur_dioxide', 'total_sulfur_dioxide', 'density', 'pH',
        'sulphates', 'alcohol','quality' ]) trained_fake_data .head()
quality_mean = trained_fake_data .quality.mean()
trained_fake_data['quality'] = trained_fake_data ['quality'] > quality_mean
trained_fake_data["quality"] = trained_fake_data ["quality"].astype(int)

features = ['fixed_acidity', 'volatile_acidity', 'citric_acid',
'residua_ sugar', 'chlorides', 'free_sulfur_dioxide', 'total_sulfur_
dioxide', 'density', 'pH', 'sulphates', 'alcohol']
label = ['quality']
x_f_predicted =trained_fake_data [features]
y_f_predicted =trained_fake_data [label]

x_f_trn, x_f_tst, y_f_trn, y_f_tst = sms.train_test_split(x_f_predicted,
y_f_predicted, random_state=99)
random_forest_fake = ensemble.RandomForestClassifier(n_estimators=200)
random_forest_fake.fit(x_f_trn,y_f_trn.values.ravel())
y_f_pred=random_forest_fake.predict(x_f_tst)
print("Fake data Accuracy ",metrics.accuracy_score(y_f_tst, y_f_pred))
print("Fake data Classification Result:",metrics.classification_report(y_f_
tst, y_f_pred))
```

Output:

Fake data Accuracy 0.965

```
Fake data Classification Result:              precision    recall
f1-score   support

           0       0.97      0.98      0.97       135
           1       0.95      0.94      0.95        65
```

```
    accuracy                          0.96      200
   macro avg       0.96      0.96      0.96      200
weighted avg       0.96      0.96      0.96      200
```

The accuracy of the fake data is 0.96 when we compare our real trained data, which is 0.76. It seems our fake data shows much better performance.

```
from table_evaluator import load_data, TableEvaluator # Table_evalutor
needs to be add to library.

evaluation_table = TableEvaluator(wine_data,trained_fake_data)
evaluation_table.evaluate(target_col='quality')
evaluation_table.visual_evaluation()
Output:
Mean Correlation between fake and real columns 0.8129
```

The absolute log mean and STDs of real and fake data are given in Figure 5-26.

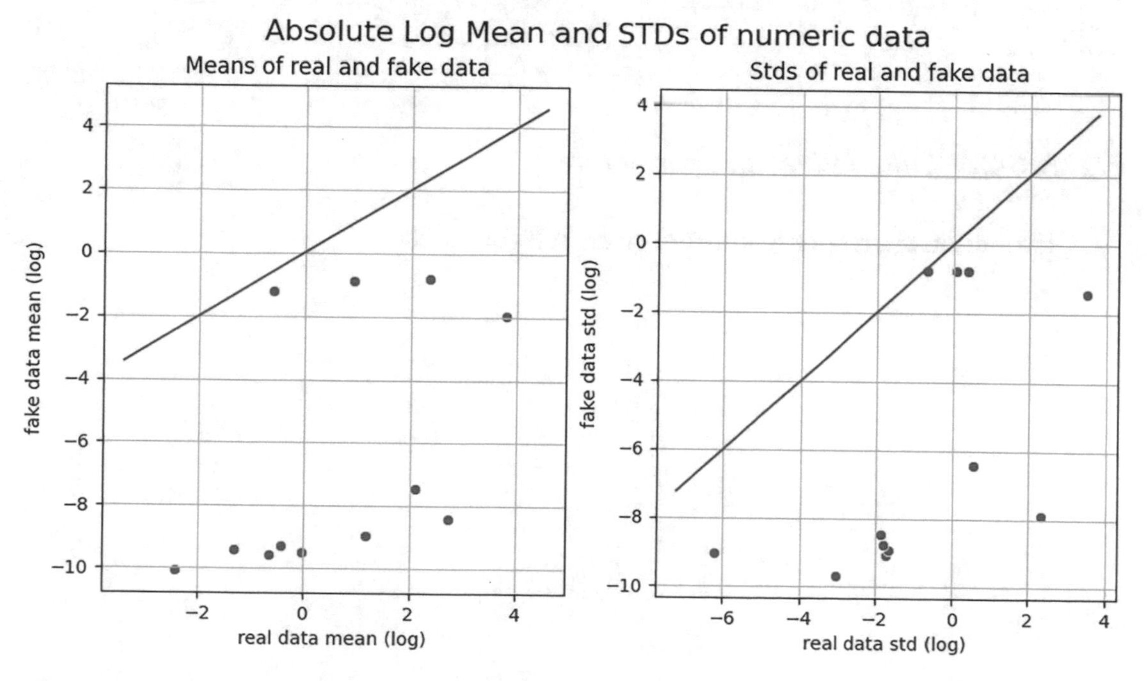

Figure 5-26. *Absolute log mean and STDs*

The correlation between the synthetic and real columns is 0.8129. It seems our synthetic data shows much better performance.

Cumulative sums according to the characteristics of the variables are given in Figure 5-27.

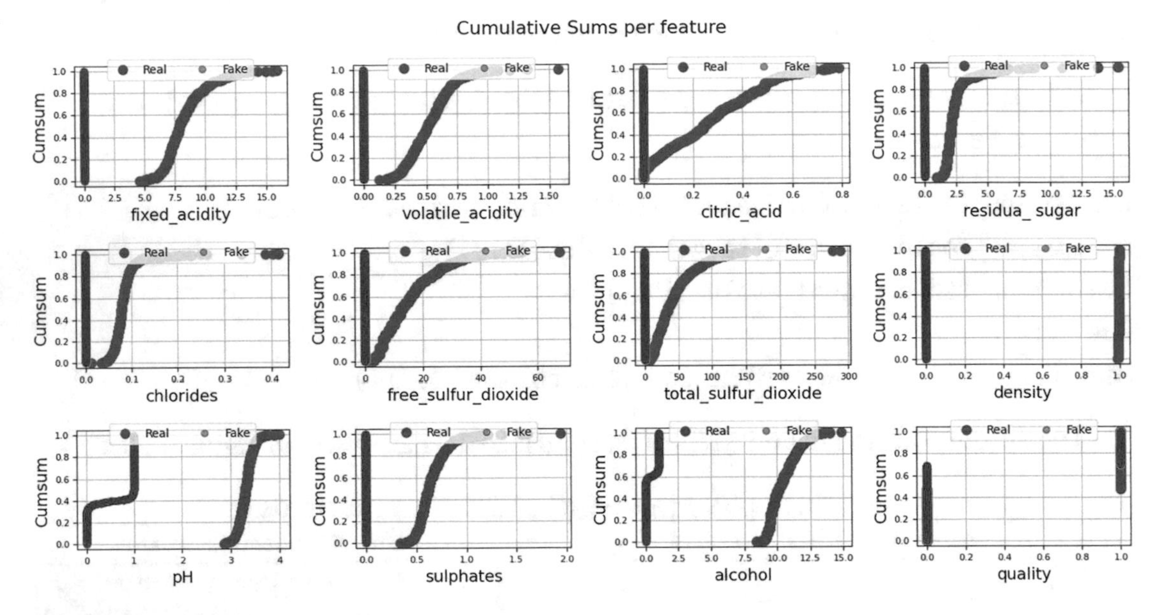

***Figure 5-27.** Cumulative sums per feature*

Distribution charts per feature are given in Figure 5-28.

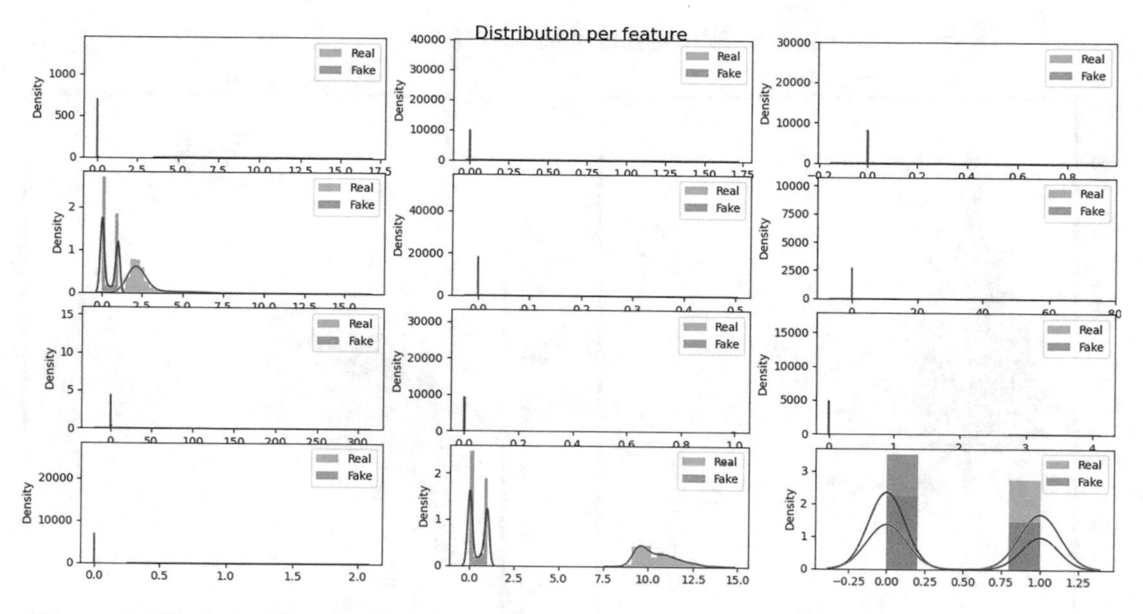

Figure 5-28. *Distribution per feature*

Figure 5-29 shows the correlation of real, fake, and difference data.

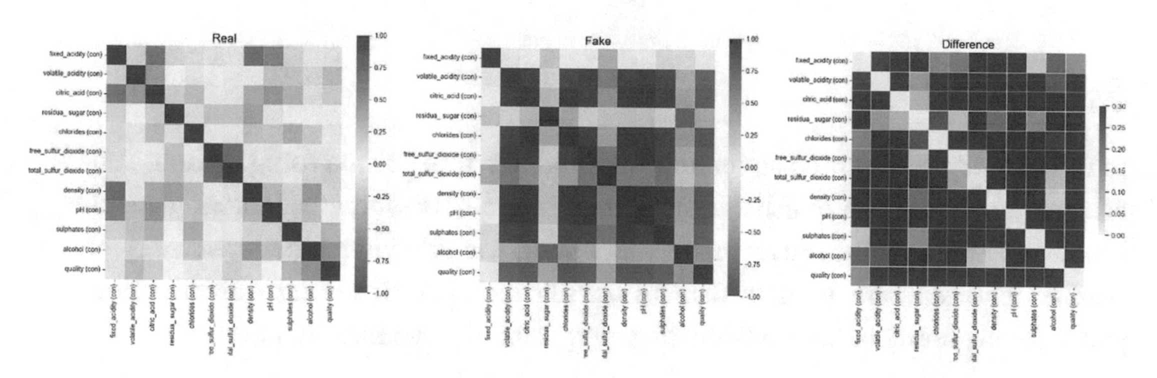

Figure 5-29. *Correlation of real, fake, and difference data*

Figure 5-30 shows the first two components of PCA.

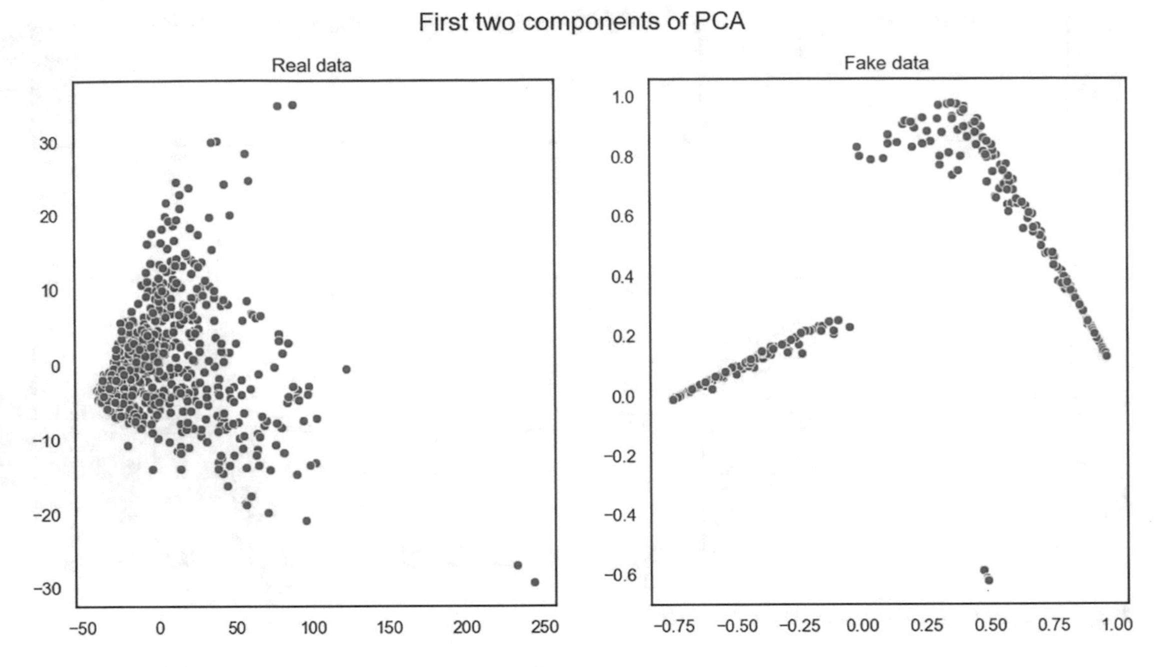

Figure 5-30. *First two components of PCA*

Summary

In this chapter, you learned about how to generate synthetic data using Python. You also learned about how to use Gaussian noise to generate synthetic data for regression problems. Additionally, you learned about how to use Friedman functions and symbolic regression to generate synthetic data for classification and clustering problems. Finally, you learned about how to use GANs to generate tabular synthetic data.

Reference

[1]. Fzhurd, "A Step -by -Step Guide to Generate Tabular Synthetic Dataset with GANs," Analytics Vidhya, 2021. `https://medium.com/analytics-vidhya/a-step-by-step-guide-to-generate-tabular-synthetic-dataset-with-gans-d55fc373c8db`.

Index

A

AlexNet, 91
Algorithmic adaptation, 46, 47
Algorithmic enhancement, 46
Anonymity, 9
Anonymization, 9
Anonymizing data, 8, 96
Archimedean copulas, 146
Artificial intelligence (AI), 3, 4, 10
Artificial neural networks, 23, 48, 84
Automotive industry, 16, 17

B

Basic functions, 75
Binomial distribution, 77
Bootstrapping, 182

C

C function, 145
Circle without noise, 200
Classification problems
  degree of difficulty, 187
  with different noise, 193
  dimensions, 183
  Multi-label, 193, 194
  noise control, 191
  parameters, 183
  samples, 185
  with 0.01 separation degree, 189
  with 5.0 separation degree, 187
  separation degree of classes, 185
class_separation parameters, 189
Clustering problems
  make_circles() function, 199
  noise levels, 202
  numbers of centers, 196, 198, 199
  randomly generated, 194
  types, 195
cluster_std feature, 198
Computer-generated imagery (CGI), 40
Computer vision, 91
  CNN, 23
  digital images, 22
  instance segmentation, 22, 23
  neural network, 23
  Object detection, 22
  segmentation problem, 26, 27
  visual scenes, 24–26
Computer vision applications, 45
Conditional GAN (cGAN), 70, 71
Conditional specification method, 104
Conjurer package in R
  customer creation, 115–117
  mathematical operations, 114
  product creation, 117
  product transactions, 118
  synthetic data generation,
    114, 119–121
Convolutional neural networks (CNNs),
    23, 47, 48

N. Gürsakal et al., *Synthetic Data for Deep Learning*, https://doi.org/10.1007/978-1-4842-8587-9

Convolutional tensor-based generative adversarial network (CTGAN), 63, 64
Copula distance graph, 157
Copulas, 145–147
CUDA, 97

D

Data augmentation, 42, 46, 47
  AlexNet, 91
  changes, 93
  computer vision, 96
  definition, 91
  GANs, 96
  image classifiers, 95
  images, 91
  NLP, 92
  Python packages, 96
  random erasing, 96
  supervised learning, 91
  transformation functions, 93, 94
Data generation techniques, 3
Data life cycle, 5–7
Data preprocessing, 43
Data privacy, 8, 9
Data quality, 10
Data science, 3, 4
Data synthesis techniques, 121
Deep convolutional generative adversarial network (DCGAN), 67, 68
Deep neural network, 49
Discrimination function, 207
Distributions, 160, 161
Domain adaptation, 44–47
Domain randomization, 48, 49
Domain transfer, 43, 44
Dual-GAN (DGAN), 63

E

Ensembles, 43

F

Facebook, 19, 52
Facial recognition model, 91
Faker, 161
Fake text data, 161
Falling Things dataset, 96
Feature adaptation, 46, 47
Friedman data generation formulas, 172
Friedman functions
  data generation formulas, 172
  dimensions and variables, 175, 177
  linear related data modeling, 172
  machine learning, 172
  predetermined, 178
  statistics and regression analysis, 172
  symbolic regression, 172, 178
Fusion data, 7
Future Importance method, 130

G

Gaussian blobs, 195, 196, 198
Gaussian copula, 147, 153–156
Gaussian distribution, 77
Gaussian mixture model, 182
Gaussian noise
  accuracy, 168
  fitted regression line, 166, 170
  regression model, 168, 169
  standard deviation, 168
Generative adversarial networks (GANs), 3
  advantages, 203
  applications, 205

autoencoders, 62
BigGAN, 71, 72
cGAN, 70, 71
CTGAN, 63, 64
data types, 203
DCGAN, 67, 68
DGAN, 63
disadvantages, 204
generator and discriminator networks, 205, 207, 210–213
libraries, 204
MedGAN, 66, 67
neural networks, 61
parts, 205
seqGAN, 69, 70
SinGAN, 66
SurfelGAN, 64, 65
tasks, 61
vs. VAEs, 63
Generative models, 32, 182
Generative networks, 32
Genset and sorter functions, 209, 210
Graphics processing unit (GPU), 97
Ground truth, 4
Gumbel copula, 146

H

Healthcare, 14, 16
Holdout data, 4

I

Image augmentation, 91
Inconsistent cluster boundaries, 195
Individual privacy, 8
Interclass distinctiveness, 96
Intraclass invariance, 96

J

Joint density, 146
Joint modeling method, 103

K

Kernel density estimation, 159
Known distribution
fake text data, 161
non-parametric model, 159
parameters, 159, 160
real data, 160
Turkish data profile, 161

L

Labeling, 9
Language models, 21
Limitless domain randomization, 41
Linear regression model, 168
List-based deletion method, 103

M

Machine learning, 18, 43–47, 54
Mahalanobis distance inequality test, 142
Mahalanobis distance ratio, 138
make_blobs() function, 195
make_circles() function, 199
make_moons() function, 201
make_regression() function, 165
Manufacturing, 12–14
Marketing, 20
Markov Chain Monte Carlo algorithm, 110
Multi-comparison model, 138
Multi-Instance Contrastive Learning, 53
Multi-label classification, 193, 194
Multivariate normal distribution, 79

N

Natural language processing (NLP), 21, 92
Neural network, 23
nnet package in R, 88
nnet package in R, neural networks
  creating and training, 84
  functions, 84
  modeling, 89
  regression model, 88
  training dataset, 87
Nodes, 97
Non-Bayesian predictive distribution, 103
Non-uniform cluster shapes, 195
Normal copula, 150–152
Normal distribution, 159

O

Object recognition, 25
Observations, 34
Overlapping clusters, 195

P, Q

Pareto principle, 119
Policy gradient, 69
Posterior predictive distribution, 103
Pretraining, 50, 51
Privacy, 9
Python library
  Faker, 161
  Scikit-learn, 183
  sklearn.dataset package, 165
  synthetic data generation, 214

R

Random erasing, 96
Real data, 7
Realistic synthetic data, 3
Real-world experience, 49, 50
Recurrent neural network (RNN), 65
Regression analysis, 164
Regression models, 168
Reinforcement learning, 7, 51, 52
Reinforcement Learning with Augmented Data (RAD), 96
Robots, 17, 18
R programming language
  basic functions, 75
  data augmentation, 91
  functions, 76
  image augmentation (*see* Torch package)
  machine learning, 90
  multivariate imputation (*see* Synthetic data imputation)
  neural networks, 84
  univariate distribution, 75

S

Security, 18
Segmentation problem, synthetic data, 26, 27
Self-driving technology, 49
Self-supervised learning, 50, 53, 54
Sentiment analysis, 92
Sequence Generative Adversarial Nets (SeqGAN), 69, 70
set.seed() function, 77
Simulations, 19, 49
sklearn.datasets library, 199

Social media, 19
Stanford AI Lab, 93
Statistical predictive modeling, 103
Statistical properties, 159
Supervised learning algorithms, 4, 91
Symbolic regression
  definition, 172
  library, 178
  samples, 180
  samples with noise, 182
SymPy library, 178
Synthetic data, 96
  accuracy problems, 4, 5
  AI systems, 3, 4, 37
  augmented data, 34
  automotive industry, 16, 17
  computer program, 1, 2
  computer vision
    CNN, 23
    digital images, 22
    input, 23
    instance segmentation, 22, 23
    neural network, 23
    object detection, 22
    segmentation problem, 26, 27
    visual scenes, 24–26
  data collection and analysis, 1
  data collection *vs.* privacy, 7
  data privacy, 8, 9
  data quality, 10
  data science, 3, 4
  fair synthetic data, 31, 32
  features, 2
  generation process, 32, 33
  healthcare, 14, 16
  importance, 2, 3
  life cycle, 5–7
  manufacturing, 12–14
  marketing, 20
  NLP, 21
  vs. real data, 48
  regression analysis, 35, 36
  regression data, 33, 34
  regression equation, 34
  regression line, 36
  robots, 17, 18
  security, 18
  social media, 19
  video games (*see* Video games)
Synthetic data imputation
  advantage, 102
  conditional specification method, 104
  definition, 102
  disadvantage, 102
  joint distribution, 104
  joint modeling method, 103
  missing values, 103, 105, 106, 108
  precision values, 103
  statistical predictive modeling, 103
Synthetic-to-real domain gap
  data augmentation, 42
  data preprocessing, 43
  definition, 42
  domain adaptation, 44–47
  domain randomization, 48, 49
  domain transfer, 43, 44
  ensembles, 43
  factors, 42
  transfer learning, 42
Synthpop package
  application, 122
  boxplot comparison, 134
  coefficient estimates, 135, 139
  confidence interval plot, 137, 141
  Copula, 145–147
  data distributions, 127

Synthpop package (*cont.*)
datasets, 122
frequency distribution, 129
Gaussian copula, 153–156
Histogram Similarity method, 127, 131
imputation methods, 122
missing values, 123
multiple comparison, 132, 133
normal copula, 150–152
synthesised and observed data, 135, 136
synthesis model, 142, 145
synthetic and observed data, 143
synthetic data distributions, 128
t copula, 147–150

T

t copula, 147–150
TF-Transformation Functions, 93
Torch package
advantages, 97
built-in nodes, 97
dimensions, 98
image augmentation, 97
image resizing, 100
interfaces, 97
machine learning library, 97
neural networks, 98
scientific computing libraries, 97
Transfer learning, 42, 45
Transformation operations, 94
Twitter, 20
Two-dimensional contour density graph, 149
Two-dimensional distribution, 149, 152

U

Uneven cluster sizes, 195
Univariate distribution, 157
Unsupervised learning, 37, 51, 54
Use_weighting function, 163

V

Value vector creation
MASS package, 79–84
multi-levels categorical variable, 78
multivariate, 78, 79
univariate distribution, 77, 78
Video games
AI, 38, 39
CGI, 40
code, 41
convincing synthetic images, 39
datasets, 39, 40
detours, 41
driverless cars, 39
Grand Theft Auto, 37
GTA V, 40
limitless domain randomization, 41
semantic segmentation, 37, 38
simulators, 39, 40
stop signs, 41
Unity, 41
Virtual environments, 3
Visual scenes, 24–26

W, X, Y, Z

Wasserstein GAN (WGAN), 68, 69

Printed in the United States
by Baker & Taylor Publisher Services